Linear and Nonlinear Wave Propagation

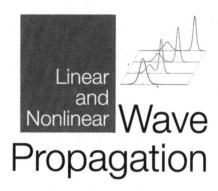

Linear and Nonlinear Wave Propagation

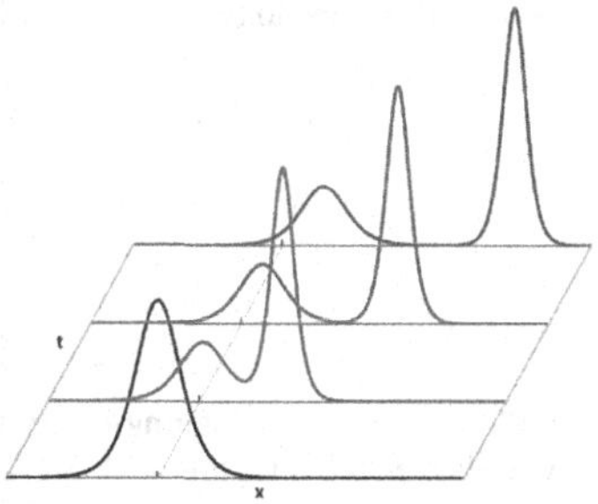

Spencer Kuo

New York University, USA

World Scientific

NEW JERSEY • LONDON • SINGAPORE • BEIJING • SHANGHAI • HONG KONG • TAIPEI • CHENNAI • TOKYO

Published by

World Scientific Publishing Co. Pte. Ltd.
5 Toh Tuck Link, Singapore 596224
USA office: 27 Warren Street, Suite 401-402, Hackensack, NJ 07601
UK office: 57 Shelton Street, Covent Garden, London WC2H 9HE

British Library Cataloguing-in-Publication Data
A catalogue record for this book is available from the British Library.

LINEAR AND NONLINEAR WAVE PROPAGATION

ISBN 978-981-123-163-6 (hardcover)
ISBN 978-981-123-164-3 (ebook for institutions)
ISBN 978-981-123-165-0 (ebook for individuals)

For any available supplementary material, please visit
https://www.worldscientific.com/worldscibooks/10.1142/12143#t=suppl

Typeset by Stallion Press
Email: enquiries@stallionpress.com

Printed in Singapore

to my parents,
Wen-Hsiu and U-Lan Kuo

and to Sophia Ping Kuo,
my wife, constant companion and best friend.
She is the loving mother to my daughter Adeline,
grandmother to Wesley Wong,
and mother-in-law to WaiKin Wong.

Preface

Waves are an essential phenomena in most scientific and engineering disciplines, such as electromagnetism and optics, and different mechanics including fluid, solid, structural, quantum, etc. The features of the waves are usually described by solutions to either linear or nonlinear partial differential equations, which are fundamental to the students and researchers.

This book was prepared to provide students with an understanding of waves and methods of solving wave propagation problems, which will enable them to expand their studies into related areas. The selection of topics and the focus given to each topic provides a way for a lecturer to cover the bases in a linear/nonlinear wave course.

Wave phenomena in linear system are introduced and discussed in the first two chapters. Generic equations describing linear wave/pulse propagation in non-dispersive and dispersive, uniform and nonuniform, with/without damping systems are analyzed as initial/boundary value problems. Methods of analyses are introduced and illustrated with analytical solutions.

Chapter 3 introduces linear wave phenomena when traversing a temporal discontinuity interface between two media. Equivalent reflection and transmission phenomena to those occurring at the spatial discontinuity interface are revealed. It illustrates the similarity and difference in the space-time duality of wave phenomena at a discontinuity interface between media.

One-dimensional lumped systems descriptive by linear and nonlinear oscillators are introduced and analyzed in Chapters 4 and 5. Mode method introduced in Chapter 2 is further elaborated via Duffing equation and exemplified with Van der Pol equation for obtaining the analytical solution. Driving a damped bistate Duffing oscillator from a deterministic state to a chaos state is illustrated. Average Lagrangian and Hamiltonian Method is introduced to open up the mode method for the quasi-harmonic responses of systems. Whitham's variational principle for the average Lagrangian is introduced to further extend the average method for periodic responses of strongly nonlinear variable parameter systems. Analytical approach for strongly nonlinear variable parameter lumped systems is also presented and the analytical solutions are used to verify the approximate responses from the average method.

In Chapter 6, Generic nonlinear equations and stationary base-band/modulated nonlinear wave solutions are illustrated. Propagation of waves in plasma is considered to mimic the nonlinear wave phenomena. Three representative nonlinear wave equations, nonlinear Schrödinger equations for EM wave and Langmuir wave, Korteweg-de Vries (K-dV) equation for ion acoustic wave, and Burgers equation for dissipated ion acoustic wave, are derived formally. Stationary properties of these nonlinear wave equations are presented in Chapter 7. In terms of the potential distributions, periodic and solitary solutions are visualized. Methods are introduced to obtain analytical solutions, which are plotted for illustration. The inverse scattering transform (IST) to solve the initial value problem on the K-dV equation is discussed and exemplified with a two-soliton problem the steps to obtain solution. The complexity of nonlinear wave behavior during the transition is illustrated with numerical results displaying the breaking of an initial Gaussian pulse into multiple non-interacting solitons. Nonlinear Burgers equation is transformed via the Cole-Hopf transformation to a linear diffusion equation.

Wave-wave and wave-particle interactions ascribed to the nonlinearity of media (such as plasma) are discussed in Chapter 8. Plasma, a Vlasov-Poisson System, instigates mode coupling, which broadens wave spectrum to lead to quasi-linear diffusion and further enhancing diffusion with resonance broadening. In spherical coordinates, the cubic nonlinearity of the nonlinear Schrödinger equation becomes inhomogeneous and evades stationary local solution. The collapse of nonlinear Langmuir wave packet is discussed.

I wish to express my sincere gratitude to Professors Bernard R. S. Cheo and Nathan Marcuvitz who guided me to this field and have given me much helpful advice and kind encouragement on my scientific evolution throughout my career. Some of the examples presented in the book are references to the class notes prepared by Professor Marcuvitz for the "Nonlinear waves" course. I would like to acknowledge Professor Min-Chang Lee for a lasting research collaboration.

Spencer Szu-Ping Kuo

Contents

List of Figures

Chapter 1

Wave Phenomena in Linear Systems

Traveling waves are the progressive motion of physical quantities which vary in space and time. Water wave is the most readily perceived example as appearing in the form of a space-time dependent disturbance moving on the water surface; on the other hand, light and sound are the two most important waves for sensing the world around us.

While the physical quantities appearing in each wave system are different, a common core of phenomena include:

1. A disturbance travels or propagates at a finite speed.
2. Energy and information are transported by the wave.
3. The wave is scattered or reflected and refracted when it encounters a change in the characteristics of the medium through which it propagates.
4. Two or more waves can interfere (add and subtract) to enhance the wave intensity in some regions and decrease it in others.

These phenomena and examples of their applications are illustrated in the following.

For wave quantities that have harmonic or sinusoidal time variation, their variation in space is also sinusoidal, with spatial period λ (called wavelength) determined by the frequency f and the phase velocity v of wave propagation. Harmonic time variation is of importance because most natural and man-made sources are nearly harmonic, usually having amplitude or frequency variations that are small over one cycle. The relation between frequency f and wavelength λ of a sinusoidal wave is given to be

$$\lambda \mathrm{f} = \mathrm{v}$$

If the phase velocity v is independent of f, the medium where the wave propagates is said to be non-dispersive, because waves with spectral bands will be shape-preserved in propagation. Electromagnetic waves propagating in free space offers the best example of non-dispersive waves. On the other hand, when wave propagates in a dispersive medium, its phase velocity v is a function of f; hence, waves with spectral bands (such as pulses) are dispersed during propagation, resulting to the change of the shape and intensity distribution.

The transmission of information, which involves energy transfer, is one of the most important engineering applications of waves. Much of our daily lives is closely involved with radio and phone communications. These communications systems employ electromagnetic carriers ranging from rf waves to millimeter waves, as well as optical waves.

If the properties of the medium through which a wave propagates is the same everywhere, the wave will continue to propagate in the same direction. However, if the wave encounters an obstacle or changes in the properties of the medium, scattering or reflection of the wave occurs, as well as changes in its direction of propagation.

When two or more waves exist in the same region, the physical quantities associated with them will add in certain parts and subtract in others. This phenomenon is called interference and is most pronounced for waves with harmonic or sinusoidal time dependence at or near the same frequency. The appearance of many waves in a system may be due to independent sources or to reflection and scattering of a single initial wave.

1.1 Wave propagation in uniform medium

Equation (1.1) represents a wave equation, whose solutions are wave functions propagating along the z axis at velocity v.

$$\left(\frac{\partial^2}{\partial t^2} - \mathrm{v}^2\frac{\partial^2}{\partial z^2}\right)U(z,t) = 0 \tag{1.1}$$

It has a general solution of the form: $U(z,t) = F(z - \mathrm{v}t) + G(z + \mathrm{v}t) = \bar{F}(z/\mathrm{v} - t) + \bar{G}(z/\mathrm{v} + t)$, where F and G (and $\bar{F}$ and $\bar{G}$) are arbitrary functions of a single variable $\mathrm{z - vt}$ and $\mathrm{z + vt}$ (and $\mathrm{z/v - t}$ and $\mathrm{z/v + t}$), respectively, and represent forward ($+\mathrm{z}$) and backward ($-\mathrm{z}$) propagating waves. Then, F and G functions (or $\bar{F}$ and $\bar{G}$) are uniquely determined by the imposed initial or boundary conditions of each specific problem.

As an example, consider the propagation of electromagnetic waves in a uniform dielectric medium with a permittivity $\epsilon = \epsilon_r\epsilon_0$, where ϵ_0 and ϵ_r are the free space permittivity and relative dielectric constant, respectively. The initial wave electric field $E_x(z, 0)$ and magnetic field $H_y(z, 0)$ are given to be

$$E_x(z, 0) = E_0 \exp(-z^2/2) \quad \text{and} \quad H_y(z, 0) = H_0 \exp(-z^2/2).$$

The intrinsic impedance of the medium $\eta = \eta_0/\sqrt{\epsilon_r}$ and the wave velocity $v = c/\sqrt{\epsilon_r}$, where the free space intrinsic impedance $\eta_0 = 120\pi$ (Ω) and c is the speed of light in free space. Hence,

$$E_x(z, 0) = \varphi_E(z, 0) = F(z) + G(z) = E_0 \exp\left(-\frac{z^2}{2}\right) \tag{1.2a}$$

and

$$H_y(z, 0) = \varphi_H(z, 0) = \frac{1}{\eta}[F(z) - G(z)] = H_0 \exp\left(-\frac{z^2}{2}\right) \tag{1.2b}$$

Equations (1.2a) and (1.2b) are solved to obtain

$$F(z) = 1/2(E_0 + \eta H_0) \exp\left(\frac{-z^2}{2}\right)$$

and

$$G(z) = 1/2(E_0 - \eta H_0) \exp\left(\frac{-z^2}{2}\right).$$

The wave fields are then determined to be

$$E_x(z, t) = 1/2\, (E_0 + \eta H_0) \exp\left[-\frac{(z - vt)^2}{2}\right] + 1/2\, (E_0 - \eta H_0) \exp\left[-\frac{(z + vt)^2}{2}\right]$$

and

$$H_y(z, t) = \frac{1}{\eta}\left\{1/2\, (E_0 + \eta H_0) \exp\left[-\frac{(z - vt)^2}{2}\right] - 1/2\, (E_0 - \eta H_0) \exp\left[-\frac{(z + vt)^2}{2}\right]\right\}$$

Next, consider an EM wave propagating from left to right along the z-axis and encountering a discontinuity at $z = 0$ between two uniform dielectric media, which have intrinsic impedances η_1 and η_2, respectively. The boundary conditions at the interface (z = 0) of the two media are given to be

$$\mathrm{E}_x(0,\mathrm{t}) = \mathrm{E}_0 \exp(-\mathrm{t}^2) \quad \text{and} \quad \mathrm{H}_y(0,\mathrm{t}) = (\mathrm{E}_0/\eta_2)\ \exp(-\mathrm{t}^2).$$

The reflection coefficient Γ and transmission coefficient τ at $z = 0$ are given to be $\Gamma = (\eta_2 - \eta_1)/(\eta_2 + \eta_1)$ and $\tau = 1 + \Gamma = 2\eta_2/(\eta_2 + \eta_1)$. The continuity conditions of the wave fields at the interface lead to

$$\begin{aligned}\mathrm{E}_x(\mathrm{z}=0^-,\mathrm{t}) &= \bar{F}_<(-\mathrm{t}) + \bar{G}_<(\mathrm{t}) = (1+\Gamma)\bar{F}_<(-\mathrm{t}) \\ &= \mathrm{E}_x(\mathrm{z}=0^+,\mathrm{t}) = \bar{F}_>(-\mathrm{t}) = \mathrm{E}_0\ \exp(-\mathrm{t}^2) \qquad (1.3\mathrm{a})\end{aligned}$$

and

$$\begin{aligned}\mathrm{H}_y(\mathrm{z}=0^-,\mathrm{t}) &= \frac{1}{\eta_1}[\bar{F}_<(-\mathrm{t}) - \bar{G}_<(\mathrm{t})] = \frac{1}{\eta_1}(1-\Gamma)\bar{F}_<(-\mathrm{t}) \\ &= \mathrm{H}_y(\mathrm{z}=0^+,\mathrm{t}) = \frac{1}{\eta_2}\bar{F}_>(-\mathrm{t}) = \frac{\mathrm{E}_0}{\eta_2}\exp(-\mathrm{t}^2). \qquad (1.3\mathrm{b})\end{aligned}$$

Equations (1.3a) and (1.3b) lead to the same results

$$\bar{F}_<(-t) = \frac{\mathrm{E}_0}{1+\Gamma}\exp(-\mathrm{t}^2), \quad \bar{G}_<(t) = \frac{\Gamma\mathrm{E}_0}{1+\Gamma}\exp(-\mathrm{t}^2),$$

and

$$\bar{F}_>(-t) = \mathrm{E}_0\exp(-\mathrm{t}^2).$$

Therefore, the wave fields in the $z < 0$ and $z > 0$ regions are determined to be

1) $z < 0$

$$\mathrm{E}_x(\mathrm{z},\mathrm{t}) = \frac{\mathrm{E}_0}{1+\Gamma}\left\{\exp\left[-\left(\frac{\mathrm{z}}{\mathrm{v}_1}-\mathrm{t}\right)^2\right] + \Gamma\exp\left[-\left(\frac{\mathrm{z}}{\mathrm{v}_1}+\mathrm{t}\right)^2\right]\right\}$$

and

$$\mathrm{H}_y(\mathrm{z},\ \mathrm{t}) = \frac{1}{\eta_1}\frac{\mathrm{E}_0}{1+\Gamma}\left\{\exp\left[-\left(\frac{\mathrm{z}}{\mathrm{v}_1}-\mathrm{t}\right)^2\right] - \Gamma\exp\left[-\left(\frac{\mathrm{z}}{\mathrm{v}_1}+\mathrm{t}\right)^2\right]\right\}$$

2) $z > 0$

$$\mathrm{E}_x(\mathrm{z},\ \mathrm{t}) = \mathrm{E}_0\ \exp\left[-\left(\frac{\mathrm{z}}{\mathrm{v}_2}-\mathrm{t}\right)^2\right]$$

and

$$H_y(z,t) = \frac{E_0}{\eta_2}\exp\left[-\left(\frac{z}{v_2}-t\right)^2\right]$$

where $v_1 = (\eta_1/\eta_0)c$ and $v_2 = (\eta_2/\eta_0)c$.

1.2 Dispersive dielectrics

Wave propagates in a dispersive medium and the propagation of each harmonic component is governed by a dispersion relation $\omega = kc/\sqrt{\epsilon_r(\omega)}$; ω and k are not linearly related hence the phase velocity $v_p = \omega/k$ of each harmonic component varies with its frequency. The phases of different harmonic components change with respect to one another in the propagation causing the dispersion of the wave and because the wave shape changes in the propagation, the wave equation is solved differently.

The essentials of the dispersive effect are implicit in the ideas of Fourier series and integrals. For simplicity, we consider scalar waves in one dimension and is governed by the dispersion relation $\omega = \omega(k) = \omega^*(-k)$, where ω is not linearly proportional to k. The scalar amplitude U(z, t) has a general solution of the integral form

$$U(z,t) = \frac{1}{4\pi}\int_{-\infty}^{\infty} A(k)\exp\{i[kz - \omega(k)t]\}dk + \text{c.c.} \tag{1.4}$$

where c.c. stands for complex conjugate and $\omega(k)$ is imposed by the dispersion relation. The spectral amplitude A(k) describes the properties of the linear superposition of different harmonic components of the wave. It is given by the Fourier transform of the initial wave amplitude U(z, 0) and its time derivative $\partial_t U(z,0)$:

$$A(k) = \int_{-\infty}^{\infty}\left[U(z,0) + \frac{i}{\omega(k)}\frac{\partial U(z,0)}{\partial t}\right]e^{-ikz}dz. \tag{1.5}$$

Substitute (1.5) into (1.4), yields

$$\begin{aligned} U(z,t) &= \frac{1}{4\pi}\iint_{-\infty}^{\infty}\left[U(z',0) + \frac{i}{\omega(k)}\frac{\partial U(z',0)}{\partial t}\right] \\ &\quad \times \exp\{i[k(z-z') - \omega(k)t]\}\,dk\,dz' + \text{c.c.} \\ &= \frac{1}{2\pi}\iint_{-\infty}^{\infty}\left[U(z',0)\cos\omega(k)t + \frac{1}{\omega(k)}\frac{\partial U(z',0)}{\partial t}\sin\omega(k)t\right] \\ &\quad \times \exp[ik(z-z')]\,dk\,dz'. \end{aligned} \tag{1.6}$$

Eq. (1.6) is an integral solution of wave propagation in a linear homogeneous dispersive medium. Explicit solutions for several initial conditions are exemplified in the following:

1. If $U(z,0) = U_0\cos k_0 z$, for all z, represents a harmonic wave, and consider two cases:

   1) $\partial_t U(z,0) = \omega(k_0)U_0 \sin k_0 z$; then

   $$A(k) = 2\pi U_0\, \delta(k - k_0)$$

   where $\delta(k - k_0)$ is a delta function; the relation $\omega(-k) = \omega(k)$ is employed. Eq. (1.4) can be integrated to obtain

   $$U(z,\ t) = U_0 \cos[k_0 z - \omega(k_0)t]$$

   This is a monochromatic traveling wave, imposed by the initial condition, which is not affected by the dispersive effect of the medium.

   2) $\partial_t U(z,0) = 0$; then

   $$A(k) = \pi U_0[\delta(k - k_0) + \delta(k + k_0)]$$

   and (1.4) is integrated to be

   $$U(z,t) = U_0 \cos k_0 z \cos \omega(k_0)t.$$

   This is a standing wave.
   One can also substitute the initial conditions into (1.6) to determine the wave function U(z, t) directly. In the general case of a highly dispersive medium or a very sharp pulse having a large spread spectrum, it makes it difficult to further proceed the integrations in (1.6); instead, a numerical approach is adopted to solve the wave equation directly to explore the behavior of the pulse propagation.
2. On the other hand, if U(z, 0) represents a finite wave train (such as a long pulse whose spatial spectrum is not too broad) or a modulated wave, the spectral amplitude A(k) will be a peaked function, which has dominant wavenumbers k_i located at the peaks of the spectrum. In this situation, the integrations in (1.6) may be carried out analytically in an approximate way; it is to expand the frequency $\omega(k)$ around each k_i for integrating over each spectral peak in (1.6). This is illustrated in the following.
   Consider the propagation of modulated Gaussian pulses with the initial conditions

1) $U(z,0) = U_0 e^{-z^2/L} \cos k_0 z$ and $\partial_t U(z,0) = \omega_0 U_0 e^{-z^2/L} \sin k_0 z$, where $\omega_0 = \omega(k_0)$.
Substitute the initial functions into (1.6), yields

$$U(z,t) = \frac{U_0}{4\pi} \iint_{-\infty}^{\infty} \left[\cos k_0 z' + i \frac{\omega_0}{\omega(k)} \sin k_0 z' \right] e^{-\frac{z'^2}{L}} \times \exp\{i[k(z-z') - \omega(k)t]\}\, dk\, dz' + \text{c.c.} \tag{1.7}$$

Now, the frequency $\omega(k)$ is expanded around k_0:

$$\omega(k) = \omega_0 + (k - k_0) \left.\frac{d\omega}{dk}\right|_{k=k_0} + \frac{1}{2}(k-k_0)^2 \left.\frac{d^2\omega}{dk^2}\right|_{k=k_0} + \cdots$$

It is then substituted into (1.7) and makes the approximation $\omega_0/\omega(k) \sim 1$, which becomes

$$U(z,t) \cong$$

$$\frac{U_0}{4\pi} e^{i(k_0 z - \omega_0 t)} \iint_{-\infty}^{\infty} e^{-\frac{z'^2}{L}} e^{-i(k-k_0)z'} e^{i[(z-v_g t)(k-k_0) - \alpha t (k-k_0)^2]}$$

$$\times\, dk\, dz + \text{c.c.} = \frac{U_0}{4\pi} e^{i(k_0 z - \omega_0 t)} \iint_{-\infty}^{\infty} e^{-\frac{1}{L}\left[z' + i\frac{L}{2}(k-k_0)\right]^2} e^{-\frac{L}{4}(k-k_0)^2}$$

$$\times\, e^{i[(z-v_g t)(k-k_0) - \alpha t (k-k_0)^2]}\, dk\, dz' + \text{c.c.} = \sqrt{\frac{L}{\pi}} \frac{U_0}{4} e^{i(k_0 z - \omega_0 t)}$$

$$\times \int_{-\infty}^{\infty} e^{-\frac{1}{4}(L + i4\alpha t)\left[k - k_0 - 2i\frac{(z-v_g t)}{(L+i4\alpha t)}\right]^2} e^{-\frac{(z-v_g t)^2}{(L+i4\alpha t)}}\, dk + \text{c.c.}$$

$$= \sqrt{\frac{L}{(L+i4\alpha t)}} \frac{U_0}{2} e^{i(k_0 z - \omega_0 t)} e^{-\frac{(z-v_g t)^2}{(L+i4\alpha t)}} + \text{c.c.}$$

$$= \frac{U_0}{[1 + (4\alpha t/L)^2]^{1/4}} e^{-\frac{L(z-v_g t)^2}{L^2 + (4\alpha t)^2}} \cos[k_0 z - \omega_0 t + \phi(z,t)].$$

where

$$v_g = d\omega/dk|_{k=k_0}, \quad \alpha = 1/2 (d^2\omega/dk^2)\,|_{k=k_0},$$

and the phase function

$$\phi(z,t) = \frac{4\alpha t (z - v_g t)^2}{[L^2 + (4\alpha t)^2]} - 1/2 \tan^{-1}\left(\frac{4\alpha t}{L}\right).$$

It shows that the shape of the wave is not preserved. The dispersive effect, $\alpha = 1/2\,(d^2\omega/dk^2)\,|_{k=k_0}$, of the medium introduces a space-time dependent phase shift $\phi(z,t)$ to the wave carrier. This modulated Gaussian pulse travels with a group velocity,

$$v_g = \left.\frac{d\omega}{dk}\right|_{k=k_0}$$

but the pulse width is spreading during propagation. The amplitude of the envelope also decreases with time, satisfying the conservation of energy:

$$\int_{-\infty}^{\infty} U(z,\,t)^2\,dz = \int_{-\infty}^{\infty} U(z,\,0)^2\,dz.$$

2) $U(z,0) = U_0 e^{-z^2/L} \cos k_0 z$ and $\partial_t U(z,0) = 0$

Eq. (1.6) leads to

$$U(z,t) = \frac{U_0}{8\pi} \iint_{-\infty}^{\infty} e^{-\frac{z'^2}{L}} \left[e^{-i(k-k_0)z'} + e^{-i(k+k_0)z'}\right] \times \exp\{i[kz - \omega(k)t]\}\; dk\, dz' + \text{c.c.} \tag{1.8}$$

Expand the frequency $\omega(k)$ around k_0 and $-k_0$ for the $e^{-i(k-k_0)z'}$ and $e^{-i(k+k_0)z'}$ terms, respectively, (1.8) is integrated to be

$$U(z,t) \cong \frac{U_0}{2[1+(4\alpha t/L)^2]^{1/4}} \left\{ e^{-\frac{L(z-v_g t)^2}{L^2+(4\alpha t)^2}} \cos[k_0 z - \omega_0 t + \phi^+(z,t)] + e^{-\frac{L(z+v_g t)^2}{L^2+(4\alpha t)^2}} \cos[k_0 z + \omega_0 t - \phi^-(z,t)] \right\}$$

where

$$\phi^{\pm}(z,t) = \frac{4\alpha t(z \mp v_g t)^2}{L^2+(4\alpha t)^2} - \frac{1}{2}\tan^{-1}\frac{4\alpha t}{L};$$

and the relation $\left.\frac{d\omega}{dk}\right|_{k=-k_0} = -\left.\frac{d\omega}{dk}\right|_{k=k_0} = -v_g$ is applied.

It represents two similar modulated Gaussian pulses propagating in opposite direction, having the same group speed and undergoing the same spreading rate.

1.3 Modes in linear systems (superposition applicable)

Sections 1.1 and 1.2 illustrate the distinction between nondispersive and dispersive wave propagation in linear media. In nondispersive homogeneous media, a wave of arbitrary amplitude and shape is capable of being propagated without distortion and with constant speed. In dispersive media, both the wave shape and wave speed are generally variable; but a harmonic wave (and only that) of arbitrary amplitude can propagate undistorted at a constant speed.

In homogeneous, stationary, and unbounded systems, the basic source free wave (mode) types have the harmonic form $\mathrm{Ae}^{-\mathrm{i}(\omega \mathrm{t}-\mathrm{kx})}$ and are termed characteristic, eigen or stationary. Each mode is characterized by a dispersion relation of the form:

$$\omega = \omega_\alpha(\mathrm{k}),$$

or

$$\mathrm{k} = \mathrm{k}_\alpha(\omega).$$

A general approach to express the solution to the equation in the integral form is Fourier transform technique. In an unbounded system with the imposed initial condition $\mathrm{U}_\alpha(\mathrm{x}, 0)$ for a mode type α, the complex solution $\mathrm{U}_\alpha(\mathrm{x}, \mathrm{t})$ is given by

$$\mathrm{U}_\alpha(\mathrm{x}, \mathrm{t}) = \frac{1}{2\pi}\int_{-\infty}^{\infty} A_\alpha(\mathrm{k})\mathrm{e}^{-\mathrm{i}[\omega_\alpha(\mathrm{k})\mathrm{t}-\mathrm{kx}]}\,\mathrm{dk} \tag{1.9a}$$

where

$$A_\alpha(\mathrm{k}) = \int_{-\infty}^{\infty} \mathrm{U}_\alpha(\mathrm{x},\, 0)\mathrm{e}^{-ikx}\,\mathrm{dx} \tag{1.9b}$$

If the imposed condition is $\mathrm{U}_\alpha(0, \mathrm{t})$, $\mathrm{U}_\alpha(\mathrm{x}, \mathrm{t})$ is given by

$$\mathrm{U}_\alpha(\mathrm{x},\, \mathrm{t}) = \frac{1}{2\pi}\int_{-\infty}^{\infty} \mathcal{F}_\alpha(\omega)\mathrm{e}^{-\mathrm{i}[\omega t-\mathrm{k}_\alpha(\omega)x]}\,\mathrm{d}\omega \tag{1.10a}$$

where

$$\mathcal{F}_\alpha(\omega) = \int_{-\infty}^{\infty} \mathrm{U}_\alpha(0, \mathrm{t})\mathrm{e}^{\mathrm{i}\omega \mathrm{t}}\,\mathrm{dt} \tag{1.10b}$$

In a bounded system (e.g., $0 \le \mathrm{x} \le \mathrm{L}$) with the imposed boundary condition $\mathrm{U}_\alpha(0, t)$ and $\mathrm{U}_\alpha(\mathrm{L}, \mathrm{t})$, Fourier transform technique is not applicable; the method of separation of variables as well as others are applied.

The wave structure (amplitude, A, frequency, ω, wavenumber, k, etc.) of each mode is constant in space and time. In the presence of a source of excitation, many such modes are excited to response to the excitation and interfere one another in the propagation. As a result, the overall wave structure in dispersive systems generally varies slowly in space and time and has an oscillatory wave-packet form, e.g., as that shown in Fig. 1.1. As described in Sec. 1.2, this wave-packet, represented by (1.9a), has a dominant wavenumber k_0 located at the peak of the spectrum $A_\alpha(k_0)$ and a corresponding frequency $\omega_0 = \omega_\alpha(k_0)$; thus it may be described in terms of a single mode type of response that is expressed to be

$$\tilde{A}_\alpha(k_0,\, x)e^{-i[\omega_\alpha(k_0)t - k_0 x]},$$

where $\tilde{A}_\alpha(k_0, x)$ represents the envelope of the wave-packet, and k_0 and $\omega_\alpha(k_0)$ are the wavenumber and frequency of the carrier of the wave-packet. $k = k(t)$ and $x = x(t)$ for a point wave-packet, form the trajectory of the point wave-packet in the (x, k) phase space.

There are basically two analytic approaches, via the modes of the system, to the determination of the response of a system to excitation:

1. The response may be determined by direct superposition of eigenmode types whose amplitudes and phases are determined by the initial excitation. This superposition requires the ability to integrate the individual mode responses over all k or ω.

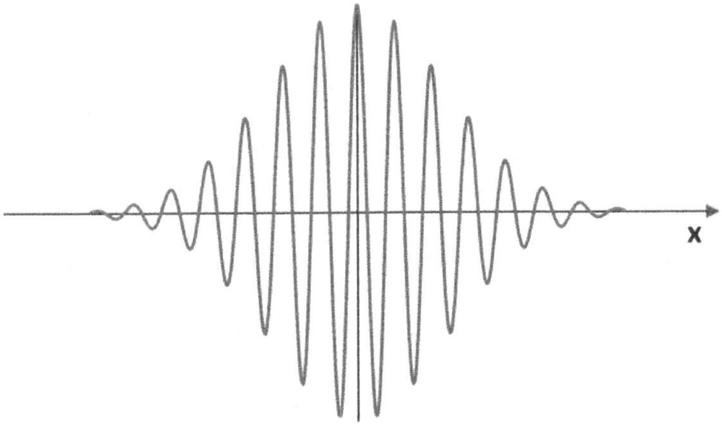

Figure 1.1. Finitely extended wave-packet response at a prescribed time.

2. In regions where wave-packets form (either propagating or non-propagating), a "ray" technique may be employed to determine at any point (x, t):

   1) a ray trajectory (x(t), k(t)) along which a point wave-packet moves, the trajectory equations, similar to the Hamilton's equations of motion, being inferred from a dispersion relation indicative of the local "frequency $\omega_\alpha(\mathrm{k})$" that represents the Hamiltonian of the ray, and
   2) the transport properties, which reveal how the local envelope and phase of the wave-packets vary along the ray trajectories.

Although superposition is still applicable for inhomogeneous or nonstationary regions of a linear system, the mode techniques begin to encounter analytic difficulties. However, the ray or wave-packet technique may still be applicable except in, so called, caustic regions, e.g., after reflection divergent rays may converge to pass a common region, where a caustic is formed; if so, they may be far more convenient to employ (in their range of applicability). Inhomogeneous or nonstationary regions give rise to a "coupling" of the waves or modes associated with the homogeneous or stationary regions.

The following are examples of simple linear systems described by source free, first order, single mode types of envelope equations in one spatial dimension, and a general three-dimensional, multi-mode types, linear envelope equation expressed in abstract form:

1. Transport equation (Non-dispersive)

$$\frac{\partial \phi}{\partial \mathrm{t}} + \mathrm{v}\frac{\partial \phi}{\partial \mathrm{x}} = 0. \tag{1.11}$$

2. Diffusion equation (Dispersive with damping)

$$\frac{\partial \phi}{\partial \mathrm{t}} + \mathrm{v}\frac{\partial \phi}{\partial \mathrm{x}} \pm (\mathrm{D} + \mathrm{iE})\frac{\partial^2 \phi}{\partial \mathrm{x}^2} = 0 \tag{1.12}$$

3. Linear Korteweg-de Vries (KdV) equation (Dispersive with no damping)

$$\frac{\partial \phi}{\partial \mathrm{t}} + \mathrm{v}\frac{\partial \phi}{\partial \mathrm{x}} \pm \mathrm{F}\frac{\partial^3 \phi}{\partial \mathrm{x}^3} = 0 \tag{1.13}$$

General multi-mode type linear envelope equation

$$L\left(\nabla, \frac{\partial}{\partial \mathrm{t}}\right) \Psi(\mathbf{r}, \mathrm{t}) = \Phi(\mathbf{r}, \mathrm{t}) \tag{1.14}$$

where L is a matrix operator and $\Psi = (\Psi_\alpha)$ and $\Phi = (\Phi_\alpha)$ are multi-component column matrices.

1.3.1 *Analytical approaches*

As illustrations, equations in 1 to 3 are solved for the initial condition

$$\phi(\mathrm{x},\, 0) = \exp\left[-\left(\frac{\mathrm{x}-\mathrm{a}}{b}\right)^2\right] \tag{1.15}$$

1) Eq. (1.11) is solved via method of characteristics, which is a general technique for solving first-order equations
The characteristic equations of (1.11) are

$$\frac{\mathrm{dx}}{\mathrm{dt}} = \mathrm{v} \quad \text{and} \quad \frac{\mathrm{d}\phi}{\mathrm{dt}} = 0 \tag{1.11a}$$

As indicated by the second characteristic equation in (1.11a), $\phi(\mathrm{x},\, \mathrm{t})$ is constant along the characteristic. The trajectory of the first characteristic in (1.11a) is given to be $\mathrm{x} = \mathrm{x}_0 + \mathrm{vt}$, where $\mathrm{x}_0 = \mathrm{x}(\mathrm{t} = 0)$ is an initial position. Thus, the solution, $\phi(\mathrm{x},\, \mathrm{t})$, of (1.11), can be constructed, with the aid of the characteristics, via the initial condition. Explicitly,

$$\phi(\mathrm{x},\, \mathrm{t}) = \phi(\mathrm{x}_0, 0) \quad \text{and} \quad \mathrm{x}_0 = \mathrm{x} - \mathrm{vt},$$

therefore, the solution of (1.11) is obtained to be

$$\phi(\mathrm{x},\, \mathrm{t}) = \phi(\mathrm{x}-\mathrm{vt},\, 0) = \exp\left[-\left(\frac{\mathrm{x}-\mathrm{vt}-\mathrm{a}}{b}\right)^2\right]. \tag{1.11b}$$

2) Eq. (1.12) is solved via the conservation condition
First, the method of characteristics is applied. The characteristic equations of (1.12) are

$$\frac{\mathrm{dx}}{\mathrm{dt}} = \mathrm{v} \quad \text{and} \quad \frac{\mathrm{d}\phi}{\mathrm{dt}} = \mp(\mathrm{D}+\mathrm{iE})\frac{\partial^2\phi}{\partial \mathrm{x}^2} \tag{1.12a}$$

Hence, along the characteristic $\mathrm{x} = \mathrm{x}_0 + \mathrm{vt}$,

$$\phi(\mathrm{x},\, \mathrm{t}) = \phi(\mathrm{x}_0 + \mathrm{vt}, \mathrm{t}) = \mathrm{V}(\mathrm{x}_0, \mathrm{t}),$$

$$\mathrm{d}\phi/\mathrm{dt} = \partial\mathrm{V}/\partial\mathrm{t} \quad \text{and} \quad \partial^2\phi/\partial\mathrm{x}^2 = \partial^2\mathrm{V}/\partial\mathrm{x}_0^2.$$

The second characteristic equation in (1.12a) becomes

$$\frac{\partial V}{\partial \mathrm{t}} \pm (\mathrm{D}+\mathrm{iE})\frac{\partial^2 \mathrm{V}}{\partial \mathrm{x}_0^2} = 0$$

In essence, this is Eq. (1.12) in a moving frame with $v_x = v$. In other words, (1.12) can be simplified to be

$$\frac{\partial\phi}{\partial t} \pm (D + iE)\frac{\partial^2\phi}{\partial x^2} = 0. \tag{1.12b}$$

In a bounded system, e.g., $0 \le x \le L$, (1.12b) is usually solved with separation of variables, subject to the two boundary conditions. However, this is an initial value problem to analyze a Gaussian pulse propagation, a different method is applied. Integration of (1.12b) along the x-axis from $-\infty$ to ∞, leads to

$$\frac{d}{dt}\int_{-\infty}^{\infty} \phi(x,\, t)\, dx = 0$$

Hence,

$$\int_{-\infty}^{\infty} \phi(x,\, t)\, dx = \text{constant} = \int_{-\infty}^{\infty} \phi(x,\, 0)\, dx$$

Since

$$\int_{-\infty}^{\infty} \exp\left[-\left(\frac{x-a}{b}\right)^2\right] dx$$

$$= \sqrt{\pi}\, b = \int_{-\infty}^{\infty} f(t)\exp\left[-\left(\frac{x-a}{b}\right)^2 f^2(t)\right] dx$$

Thus, preset

$$u(x,\, t) = f(t)\exp\left[-\left(\frac{x-a}{b}\right)^2 f^2(t)\right]$$

where $f(t)$ is to be defined and $f(0) = 1$. Substitute it into (1.12b), an equation for $f(t)$ is derived to be

$$\frac{df(t)}{dt} = \pm 2\frac{(D + iE)}{b^2} f^3(t)$$

It is solved to obtain

$$f(t) = \frac{1}{\sqrt{1 \mp 4\frac{(D+iE)}{b^2}t}}$$

Therefore, the solution of (1.12b) is obtained to be

$$\phi(x,\, t) = \frac{1}{\sqrt{1 \mp 4\frac{(D+iE)}{b^2}t}} \exp\left[-\frac{\left(\frac{x-a}{b}\right)^2}{1 \mp 4\frac{(D+iE)}{b^2}t}\right] \tag{1.12c}$$

3) Eq. (1.13) is solved via Fourier transform
First, the method of characteristics is applied to transform (1.13) to a diffusion equation

$$\frac{\partial \phi}{\partial t} = \upsilon \frac{\partial^3 \phi}{\partial x^3} \tag{1.13a}$$

where the diffusion coefficient $\upsilon = \mp F$. Eq. (1.13a) has the dispersion relation $\omega = \upsilon k^3$; for a Fourier component with initial condition $\mathcal{P}(x, 0) = e^{ikx}$, (1.13a) has solution $\mathcal{P}(x, t) = e^{i(kx-\upsilon k^3 t)}$. Since the initial condition (1.15) has Fourier spectrum

$$A(k) = \sqrt{\pi} b \exp\left[-(bk/2)^2\right] \exp(-ika)$$

thus, the solution to (1.13a) is constructed to be

$$\phi(x, t) = \frac{b}{2\sqrt{\pi}} \int_{-\infty}^{\infty} e^{-\left(\frac{bk}{2}\right)^2} e^{i[k(x-a)-\upsilon k^3 t]}\, dk \tag{1.13b}$$

However, this integral solution is implicit; (1.13a) will be solved numerically. Two conservation conditions, on (1.13a), are

$$\int_{-\infty}^{\infty} \phi(x, t) dx = \text{constant} = b\sqrt{\pi}$$

and

$$\int_{-\infty}^{\infty} \phi^2(x, t) dx = \text{constant} = b\sqrt{\pi/2},$$

which are imposed in the numerical analysis.
Equations in (1.11) to (1.13) are represented by a single equation of a general form:

$$\frac{\partial \phi}{\partial t} + v\frac{\partial \phi}{\partial x} \pm (D + iE)\frac{\partial^2 \phi}{\partial x^2} \pm F\frac{\partial^3 \phi}{\partial x^3} = 0 \tag{1.16}$$

where the coefficients D, E and F of (1.16) are set properly to match each equation in (1.11) to (1.13). Eq. (1.16) is solved for the initial condition

$$\phi(x, 0) = e^{-x^2}$$

in the cases:

(a) $v = 1$, and $D = E = F = 0$; corresponding to Eq. (1.11);
(b) $v = 0$, $D = \mp 0.5$, and $E = F = 0$; corresponding to Eq. (1.12b);
(c) $v = 0, E = \mp 0.5$, and $D = F = 0$; corresponding to Eq. (1.12b); and
(d) $v = 0, D = E = 0$, and $F = \mp 0.1$; corresponding to Eq. (1.13a) with $\upsilon = 0.1$.

The analytical solutions in the cases of (a) to (c) are

(a) $\phi(x, t) = \exp[-(x - t)^2]$;
(b) $\phi(x, t) = \frac{1}{\sqrt{1+2t}}\exp\left(-\frac{x^2}{1+2t}\right)$; and
(c) $\phi(x, t) = \phi_r(x, t) + i\phi_i(x, t) = \frac{1}{\sqrt{1+i2t}}\exp(-\frac{x^2}{1+i2t})$;
where $\phi_r(x, t) = p(t)^{\frac{1}{4}}\cos[(1 - 2x^2p)t]\exp[-x^2p(t)]$, $\phi_i(x, t) = -p(t)^{\frac{1}{4}}\sin[(1 - 2x^2p)t]\exp[-x^2p(t)]$, and $p(t) = 1/(1 + 4t^2)$.

In the case (d), Eq. (1.13a) is solved numerically.

1.3.2 *Numerical display*

Typical numerical displays of $\phi(x, t)$, in the 4 cases, as a function of x for various t are shown in Figures 1.2a–d.

1. In Fig. 1.2a, the pulse propagates in a non-dispersive medium; as shown, the pulse is undistorted and propagates at a constant velocity $v = 1$.
2. In Fig. 1.2b, the pulse propagates in a lossless medium with dispersive damping on the modes; as shown, the pulse is spreading in time and lowering the amplitude; but the area of the pulse is conserved.
3. In Fig. 1.2c, the pulse propagates in a dispersive medium; the dispersion introduces even (in k) phase shifts in time among the modes, which causes the pulse to spread in time and form symmetric spatial oscillation. The net area of the pulse is conserved, i.e., the area under $\phi_r(x, t)$ (the area above the x-axis subtracts the area below the x-axis) is conserved and the area under $\phi_i(x, t)$ is zero.
4. In Fig. 1.2d, pulse propagates in a dispersive medium; however, the dispersion introduces odd (in k) phase shifts among the modes, it causes the pulse to spread in time and form asymmetric spatial oscillation. As, k and –k modes are conjugate to each other, $\phi_i(x, t) = 0$. Areas under $\phi(x, t)$ and $[\partial\phi(x, t)/\partial x]^2$ are conserved.

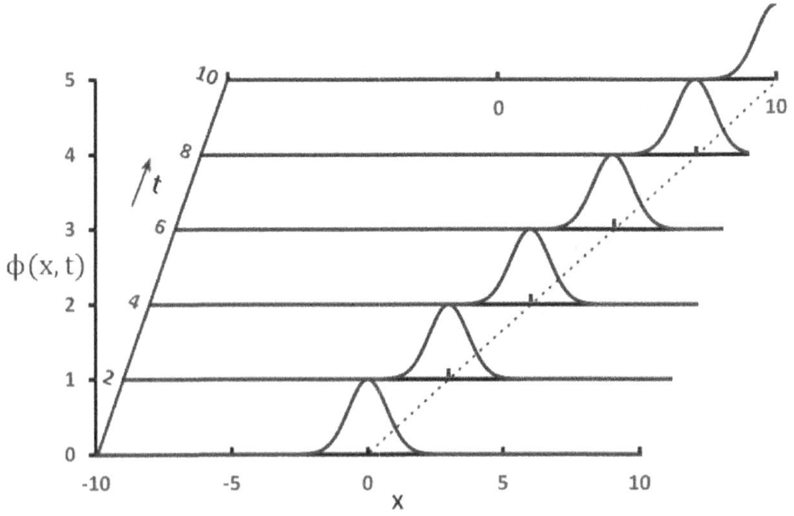

(a) Linear, non-dispersive wave solution.

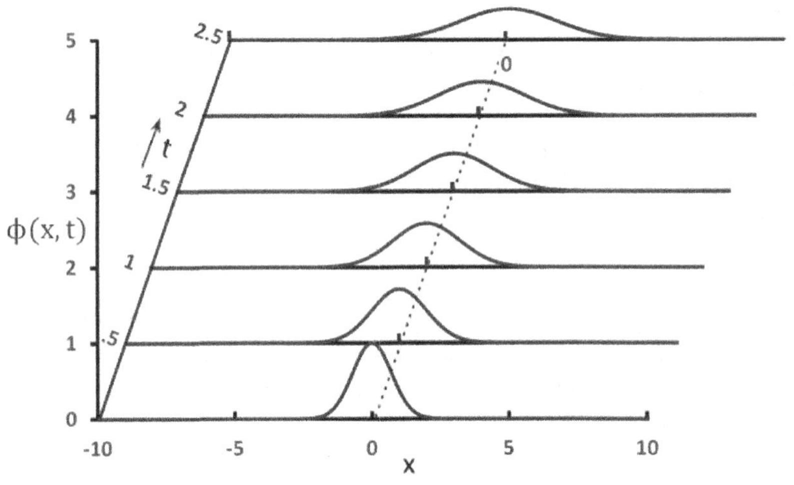

(b) Linear diffusion solution.

Figure 1.2. Linear waves.

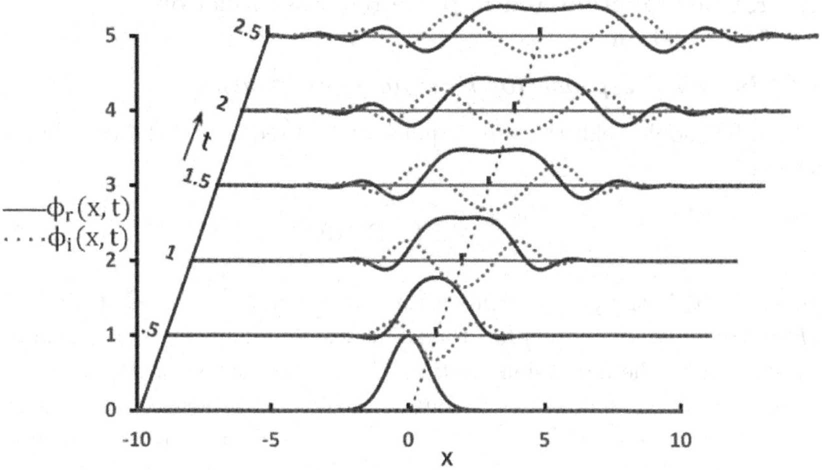

(c) Linear, dispersive wave solution.

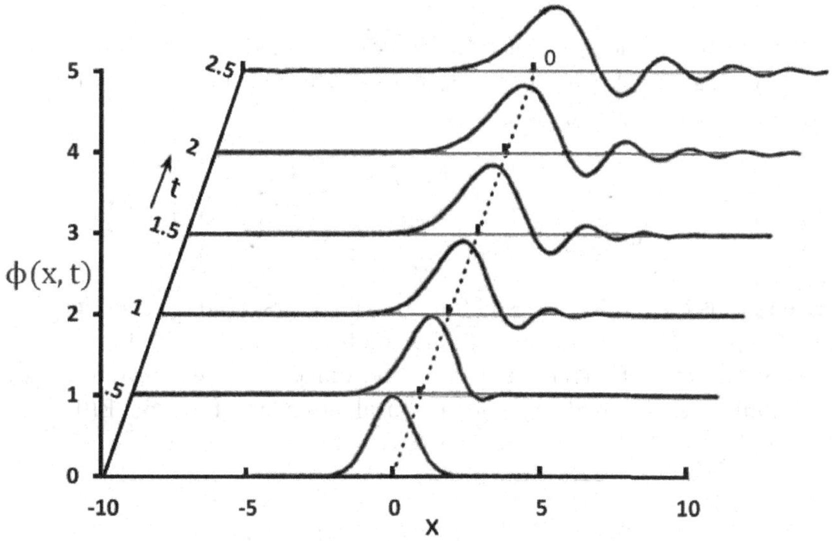

(d) Linear, dispersive wave solution (Airy type equation).

Figure 1.2. (*Continued*)

1.4 Transfer function and impulse response function of the system

1.4.1 *Impulse response for time harmonic pulses*

For time harmonic field, the wave equation (1.1) reduces to the Helmholtz equation

$$\frac{d^2E}{dz^2} + k^2(\omega)E = 0 \tag{1.17}$$

where $E(z, \omega)$ is the phasor field of the wave field $\hat{E}(z, t)$, i.e., $\hat{E}(z,t) = R_e[E(z, \omega)e^{-i\omega t}]$. Eq. (1.17) has the solutions $E^{\pm}(z, \omega) = e^{\pm ik(\omega)z}$, where $k(\omega)$ is given by the dispersion relation of the mode of the medium.

For a special group of dispersive media that $k(\omega) = \frac{i}{c}\sqrt{(-i\omega + A)^2 - B^2}$, where A and B are constants, and c is the speed of light in vacuum, the phasor of a time harmonic solution

$$H(z, \omega) = e^{ik(\omega)z} = \exp\left(-\sqrt{(-i\omega + A)^2 - B^2}\frac{z}{c}\right) \tag{1.18}$$

is assumed to be the spectral transfer function of the medium, then the corresponding impulse response function of the medium is given by the inverse Fourier transform of (1.18):

$$h(z,t) = \delta\left(t - \frac{z}{c}\right)e^{-A\frac{z}{c}} + B\frac{z}{c}\frac{I_1\left(B\sqrt{t^2 - \left(\frac{z}{c}\right)^2}\right)}{\sqrt{t^2 - \left(\frac{z}{c}\right)^2}}e^{-At}u\left(t - \frac{z}{c}\right) \tag{1.19}$$

where $I_1(\varsigma)$ is the modified Bessel function of the first kind of order one; the unit step $u(t - \frac{z}{c})$ enforces the causality condition: $h(z, 0) = 0$, for $t < z/c$.

The solution of (1.17) is then determined by the convolution of the impulse response $h(z,t)$ and a causal boundary function $\hat{E}(0,t) = \mathscr{E}(0,t)u(t)$ to be

$$\hat{E}(z, t) = \int_{-\infty}^{\infty} h(z, t')\hat{E}(0, t - t')\,dt' \tag{1.20}$$

As an example, consider pulse propagation in lossless plasma by setting $A = 0$ and $B = -i\omega_p$ in (1.19), it becomes

$$h(z,t) = \delta\left(t - \frac{z}{c}\right) - \omega_p\left(\frac{z}{c}\right)\frac{J_1\left(\omega_p\sqrt{t^2 - \left(\frac{z}{c}\right)^2}\right)}{\sqrt{t^2 - \left(\frac{z}{c}\right)^2}}u\left(t - \frac{z}{c}\right) \tag{1.21}$$

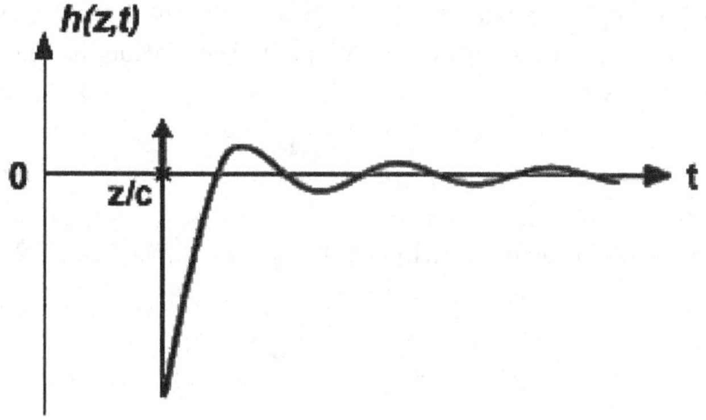

Figure 1.3. Impulse response function (1.21).

where the relation $I_1(-ix) = -i\,\mathrm{J}_1(x)$, with $\mathrm{J}_1(x)$ being the ordinary Bessel function, is adopted. A plot of $h(\mathrm{z},\ \mathrm{t})$ for $\mathrm{t} \geq z/c$ is shown in Fig. 1.3. Substitute (1.21) into (1.20), yields

$$\hat{\mathrm{E}}(z,t) = \mathscr{E}\left(0, t - \frac{z}{c}\right)\mathrm{u}\left(\mathrm{t} - \frac{z}{c}\right) - \omega_p\left(\frac{z}{c}\right)\int_{\frac{z}{c}}^{t} \frac{\mathrm{J}_1\left(\omega_p\sqrt{t'^2 - \left(\frac{z}{c}\right)^2}\right)}{\sqrt{t'^2 - \left(\frac{z}{c}\right)^2}}\mathscr{E}(0, \mathrm{t} - t')\mathrm{u}(\mathrm{t} - t')dt'. \tag{1.22}$$

1.4.2 *Impulse response for ultra-short pulses*

Consider a pulse, $E(z,t) = A(z,t)exp[i(k_0 z - \omega_0 t)]$, propagation in a dispersive medium, usually, the first step is to remove the carrier of the pulse from the wave equation to obtain an equation describing the evolution of the pulse envelope A(z, t). In the process, approximation and variable transform are made to simplify the resulting equation. When the pulse is wide in time, the second time derivative term $(\partial^2/\partial t^2)$ is neglected and spatial coordinate z is transformed to z′, in a moving frame at the phase velocity, $\mathrm{v_p} = \omega_0/\mathrm{k_0}$, of the carrier (i.e., $\mathrm{t}' = \mathrm{t}$ and $\mathrm{z}' = \mathrm{z} - \mathrm{v_p t}$); on the other hand, the second spatial derivative term $(\partial^2/\partial z^2)$ is neglected when the pulse is very short in time, and a delay time t′ is introduced to simplify the envelope equation (i.e., $\mathrm{t}' = \mathrm{t} - \mathrm{z}/\mathrm{v_p}$ and $\mathrm{z}' = \mathrm{z}$). Thus, the envelope A(z, t) of a very short pulse obeys the following equation:

$$i\frac{\partial A(z,t)}{\partial z} = \frac{\beta_2}{2}\frac{\partial^2 A(z,t)}{\partial t^2} \tag{1.23}$$

where (z', t') is replaced back to (z, t) to simply the presentation; $\beta_2 = n(\omega_0)/\omega_0 c$ and $n(\omega_0)$ is the refractive index. This equation has the solutions $A(z,\omega) = e^{ik(\omega)z}$, where

$$k(\omega) = \frac{\beta_2\omega^2}{2}.$$

Hence, the spectral transfer function of the system defined by (1.23) is given to be

$$H(z,\omega) = \exp\left(i\frac{\beta_2\omega^2}{2}z\right). \tag{1.24}$$

Then the corresponding impulse response function of the system is given by the inverse Fourier transform of (1.24):

$$h(z,t) = (-i2\pi\beta_2 z)^{-1/2} \exp\left(-i\frac{t^2}{2\beta_2 z}\right). \tag{1.25}$$

If A(0, t) is the starting pulse profile, then the pulse profile A(z, t), after propagating in the medium to a location z, can be evaluated by the following convolution:

$$A(z,t) = \int_{-\infty}^{\infty} h(z,t-t')A(0,t')\, dt'. \tag{1.26}$$

As an example, consider the propagation of a Gaussian pulse which starts at $z = 0$ with the time profile

$$A(0,t) = A_0 \exp\left(-\frac{t^2}{\tau^2}\right). \tag{1.27}$$

Substitute (1.27) into (1.26) with the aid of (1.25), yields

$$A(z,t) = \frac{A_0\tau}{\sqrt{\tau^2 - i2\beta_2 z}} \exp\left(-\frac{t^2}{\tau^2 - i2\beta_2 z}\right). \tag{1.28}$$

Dispersion causes the pulse lowering amplitude to spread along z.

Problems

P1.1. Apply coordinate transform: $\xi = z - vt$ and $\eta = z + vt$ on Eq. (1.1), show that the transformed equation can be integrated directly to obtain the general solution $\varphi(z,t) = F(z - vt) + G(z + vt)$.

P1.2. For initial value problems with the initial conditions: $\mathrm{U}(z,0)$ and $\frac{\partial \mathrm{U}(z,0)}{\partial t}$ given, the solution of the wave equation (1.1) can be expressed in terms of the initial conditions by first taking the Laplace transform of Eq. (1.1). Show that the Laplace transform function $\mathrm{L}(z, \omega)$ satisfies the equation

$$\frac{\partial^2}{\partial z^2}\mathrm{L}(z,\omega) + \frac{\omega^2}{v^2}\mathrm{L}(z,\omega) = \frac{i\omega}{v^2}\left[\mathrm{U}(z,0) + \frac{i}{\omega}\frac{\partial}{\partial t}\mathrm{U}(z,0)\right] \qquad \text{(P1.1)}$$

where

$$\mathrm{L}(z,\omega) = \int_0^\infty \mathrm{U}(z,t)\exp(i\omega t)\,dt, \ \mathrm{Im}(\omega) > 0. \qquad \text{(P1.2)}$$

P1.3. For boundary value problems with the boundary conditions: $\mathrm{U}(0,t)$ and $\frac{\partial \mathrm{U}(0,t)}{\partial z}$ given, Eq. (1.4) will be re-expressed to be

$$\mathrm{U}(z,t) = \frac{1}{4\pi}\int_{-\infty}^{\infty} \mathcal{F}(\omega)\exp\{i[k(\omega)z - \omega t]\}\,d\omega + \text{c.c.} \qquad \text{(P1.3)}$$

Show that $\mathcal{F}(\omega)$ is given by the Fourier transform of the wave amplitude $\mathrm{U}(0, t)$ and its spatial derivative $\partial_z \mathrm{U}(0,t)$ at the boundary $z = 0$:

$$\mathcal{F}(\omega) = \int_{-\infty}^{\infty}\left[\mathrm{U}(0,t) - \frac{i}{k(\omega)}\frac{\partial \mathrm{U}(0,t)}{\partial z}\right]e^{i\omega t}\,dt \qquad \text{(P1.4)}$$

and hence

$$\mathrm{U}(z,t) = \frac{1}{2\pi}\iint_{-\infty}^{\infty}\left[\mathrm{U}(0,t')\cos k(\omega)z - \frac{1}{k(\omega)}\frac{\partial \mathrm{U}(0,t')}{\partial z}\sin k(\omega)z\right] \times \exp[i\omega(t-t')]\,d\omega\,dt' \qquad \text{(P1.5)}$$

P1.4 Consider the propagation of a Gaussian modulated oscillation in a dispersive medium, where the dispersion relation of the wave is $\omega(k) = \omega_0(1 + \alpha k^2)$; the initial conditions of the pulse are

$$u(x,0) = 2e^{-x^2/2L^2}\sin k_0 x \quad \text{and} \quad \frac{\partial}{\partial t}u(x,0) = 0$$

Find $u(x,t)$.

P1.5. Verify the analytical solutions in the cases of (b) and (c) for Eq. (1.16).

P1.6. A time harmonic wave, having a causal boundary condition $\hat{E}(0,t) = \tilde{A}(0,t)e^{-i\omega t}\mathrm{u}(\mathrm{t})$, propagates in three different media: 1. A = 0 = B, representing vacuum or an ideal dielectric; 2. A > 0 and B = 0, representing a weakly conducting dielectric; and 3. A = B > 0, representing medium with finite conductivity σ, where A and B are the parameters in Eq. (1.18). Find $\hat{\mathrm{E}}(z,t)$ in three media.

P1.7. Consider the propagation of an ultra-short pulse in plasma; the pulse has a carrier frequency ω_0, and a rectangular envelope $A(0,t) = A_0[\mathrm{u}(\mathrm{t})-\mathrm{u}(\mathrm{t}-\mathrm{t}_0)]$ at the starting point z = 0, where t_0 is the pulse temporal width. The plasma has a relative dielectric function $\varepsilon(\omega) = 1 - \frac{\omega_p^2}{\omega^2} = \left(\frac{kc}{\omega}\right)^2$. Find and plot A(z, t).

Chapter 2

Wave Propagation in Linear Inhomogeneous Media

2.1 WKB solution

Inhomogeneity of the medium complicates the propagation of wave because the wavelength of each harmonic component varies continuously in the propagation. A plane wave propagates in a weakly inhomogeneous plasma is first considered to illustrate the Wentzel–Kramers–Brillouin (WKB) solution.

Consider vertical propagation of a plane wave of radian frequency ω_0 along the density inhomogeneity (z axis) of the ionospheric plasma, the wave field is governed by the Helmholtz equation

$$\frac{d^2E}{dz^2} + k^2(z)E = 0 \tag{2.1}$$

where

$$k(z) = \frac{\left[\omega_0^2 - \omega_p^2(z)\right]^{\frac{1}{2}}}{c} \quad \text{and} \quad \omega_p(z) = \left[\frac{n_e(z)e^2}{m_e\epsilon_0}\right]^{\frac{1}{2}}$$

are the wavenumber and the electron plasma frequency, respectively — geomagnetic field is not included in the formulation to simplify the analysis and presentation — the electron density $n_e(z)$ increases with the altitude, thus k(z) decreases as wave propagates upward. Consider the operation that the wave frequency $f_0 = \omega_0/2\pi$ is less than foF2 of the ionosphere (where the plasma density is the maximum). In this case, the wave can reach a layer at $z = z_0$, where $\omega_p(z_0) = \omega_0$ and $k(z_0) = 0$. This point z_0 is called the turning point, where wave reflection occurs. If the wave is a distance away below the reflection layer, WKB solution of (2.1), which is to be shown, is a good approximation.

Set

$$E(z) = A(z) \exp\left[\pm i \int k(z)dz\right],$$

where A(z) is a slow varying amplitude so that $d^2A(z)/dz^2 \sim 0$ is assumed, and substitute it into (1), yields

$$2k\frac{dA}{dz} + A\frac{dk}{dz} = 0.$$

It is then integrated to be $A(z) = A_\pm/k^{1/2}$; thus, the WKB solution is obtained as

$$E_\pm(z) = \frac{A_\pm}{k^{\frac{1}{2}}} \exp\left[\pm i \int k(z)dz\right] \tag{2.2}$$

where "±" stand for upward (forward) and downward (backward) propagation. The phasor fields of (2.2) are then converted to the real solution of (2.1):

$$\begin{aligned} E(z, t) = {} & \frac{A_0}{k^{\frac{1}{2}}} \cos\left[\int k(z)dz - \omega_0 t\right] \\ & + \frac{|A_-|}{k^{\frac{1}{2}}} \cos\left[\int k(z)dz + \omega_0 t - \theta_-\right] \end{aligned} \tag{2.3}$$

where $A_0 = A_+$ is assumed to be real and $A_- = |A_-| e^{i\theta_-}$.

However, to be consistent with the assumption that $d^2A(z)/dz^2 \sim 0$, which, in essence, is a forward scattering approximation to be shown in the following; the backward propagating component in (2.3) should be neglected and the WKB solution of (2.1) for a forward propagating plane wave is given to be

$$E(z,t) = \frac{E_i}{(k/k_i)^{1/2}} \cos\left[\int k(z)dz - \omega_0 t\right]$$

where E_i and k_i are the initial amplitude and wavenumber of the incident wave at $z = z_i$. The wave magnetic field $H(z,t) = E(z,t)/\eta$, where the intrinsic impedance $\eta = \eta_i(n_i/n)$; η_i and n_i and $n = n(z) = (k/k_i)n_i$ are the intrinsic impedance and the refractive index of the medium at $z = z_i$, and the refractive index of the medium at z. Hence, the time average Poynting vector $\langle \mathbf{S} \rangle = \langle \mathbf{E} \times \mathbf{H} \rangle = E_i^2/2\eta_i =$ constant. It indicates that the reflection has been neglected in the WKB approximation. In other words, WKB method considers only forward scattering, which is a good assumption when the spatial variation of the medium is smooth. As the wave approaches the reflection layer, the forward scattering approximation does not hold any more. Hence, the WKB solution is not appropriate, a full wave analysis on (2.1) is necessary.

In a particular inhomogeneity profile, the WKB solution turns out to be exact. Assume that the plasma density distribution from the ground to region below the reflection height z_0, i.e., $0 \leq z < z_0$, is modelled as

$$n_e(z) = n_0 \left[1 - \frac{1}{(z/z_0 + 1)^4}\right],$$

where $n_0 = (m_e \epsilon_0 / e^2)\omega_0^2$.

Hence,

$$k^2(z) = \frac{k_0^2}{(z/z_0 + 1)^4} \quad \text{for} \quad z \geq 0,$$

where $k_0 = \omega_0/c$ is free space wavenumber of a plane wave at frequency ω_0

The Helmholtz equation (2.1) becomes

$$\frac{d^2E}{dz^2} + \left[k_0^2/(z/z_0 + 1)^4\right] E = 0. \tag{2.1a}$$

The WKB solution (2.2) becomes

$$E(z) = A_0 \left(1 + \frac{z}{z_0}\right) \exp\left[i\left(1 - \frac{1}{1 + z/z_0}\right) k_0 z_0\right]. \tag{2.2a}$$

It turns out that (2.2a) is an exact solution to the equation (2.1a) because $d^2A(z)/dz^2 = 0$.

2.2 Solution of the wave equation near a turning point

In the vicinity of $k \to 0$, the electron density can be assumed to have a linearly increasing profile, i.e.,

$$n_e(z) = n_0 \left[1 + \frac{z - z_0}{L}\right],$$

where $n_0 = n_e(z_0)$ and L is the linear scale length.

Hence,

$$\omega_p^2(z) = \omega_0^2\left[1 + \frac{z - z_0}{L}\right] \quad \text{and} \quad k^2(z) = \frac{[\omega_0^2 - \omega_p^2(z)]}{c^2} = \nu(z_0 - z),$$

where $\nu = \omega_0^2/Lc^2$. A new coordinate $g = z_0 - z$ is introduced to re-express (2.1) to be

$$\frac{d^2E}{dg^2} + \nu g\, E = 0. \tag{2.4}$$

a coordinate transformation, $y = (2/3)(\nu g^3)^{1/2}$, it leads to

$$\frac{d}{dg} \to \left(\frac{3\nu y}{2}\right)^{\frac{1}{3}} \frac{d}{dy}$$

and

$$\frac{d^2}{dg^2} \to \left(\frac{3\nu y}{2}\right)^{\frac{2}{3}} \frac{d^2}{dy^2} + (\nu/2)\left(\frac{3\nu y}{2}\right)^{-\frac{1}{3}} \frac{d}{dy}.$$

Introduce another transformation by setting $E = g^{1/2}F$, i.e., $E = \nu^{-1/2}(3\nu\, y/2)^{1/3}F$, (2.4) is transformed to a Bessel equation

$$\frac{d^2F}{dy^2} + y^{-1}\frac{dF}{dy} + \left(1 - \frac{1}{9y^2}\right) F = 0. \tag{2.5}$$

The solution of (2.5) is a Bessel function of order 1/3, i.e., $F = J_{\pm 1/3}(y)$, for $y^2 > 0$, or a modified Bessel function of the second kind of order 1/3, i.e., $F = K_{1/3}(|y|)$, for $y^2 < 0$.

Therefore, the solutions of (2.4) in the two regions $g > 0$ and $g < 0$ are found to be

$$E_\pm = A_\pm g^{\frac{1}{2}} J_{\pm 1/3}\left(\frac{2}{3}\sqrt{\nu} g^{\frac{3}{2}}\right) \quad \text{for} \quad g > 0$$

and

$$E = D\, |g|^{\frac{1}{2}}\, K_{1/3}\left(\frac{2}{3}\sqrt{\nu}\, |g|^{\frac{3}{2}}\right) \quad \text{for} \quad g < 0. \tag{2.6}$$

Apply the continuity condition to (2.6) at $g = 0$, i.e., at $z = z_0$, the solution of (2.4) in the region around $z = z_0$ is obtained to be

$$E = \begin{cases} E_0 g^{\frac{1}{2}} \left[J_{1/3}\left(\frac{2}{3}\sqrt{\nu} g^{\frac{3}{2}}\right) + J_{-1/3}\left(\frac{2}{3}\sqrt{\nu} g^{\frac{3}{2}}\right)\right] & \text{for} \quad z \le z_0 \\ E_+ |g|^{\frac{1}{2}} K_{1/3}\left(\frac{2}{3}\sqrt{\nu}\, |g|^{\frac{3}{2}}\right) & \text{for} \quad z > z_0 \end{cases} \tag{2.7}$$

where $E_+ = E_0\left[J_{1/3}(0) + J_{-1/3}(0)\right]/K_{1/3}(0)$.

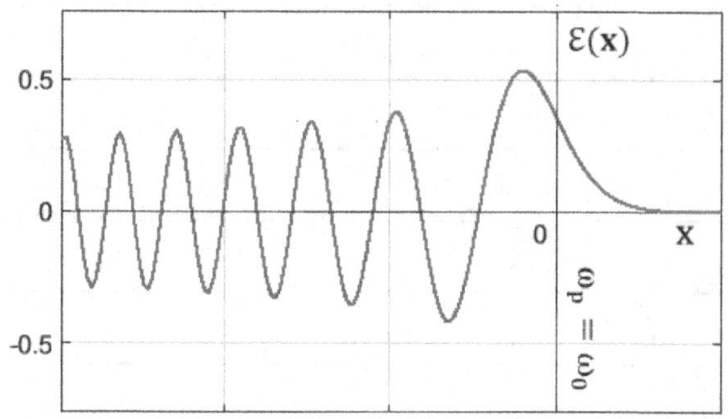

Figure 2.1. Distribution of the wave electric field near the turning point, where $x = \nu^{1/3}z$.

Set the turning point $z_0 = 0$, thus $g = -z$. Introduce the dimensionless coordinate $x = \nu^{1/3}z = \left(\omega_0^2/Lc^2\right)^{1/3} z$, the normalized field function $\mathscr{E}(x) = E/E_0$ of (2.7) is plotted in Fig. 2.1. As shown, the field amplitude near the turning point (i.e., the reflection height) is enhanced approximately by a factor of 4. This is called "swelling effect".

The wave group velocity is $v_g = c\sqrt{1 - \omega_p^2(z)/\omega_0^2}$; thus, the wave slows down while it is approaching the reflection layer at $z = 0$, where $\omega_p^2(0) = \omega_0^2$ and $v_g = 0$. The continuity condition of the power flow forces the wave energy density to buildup, 'introducing' a field swelling factor of ~2. The total reflection at cutoff adds another multiplication factor of ~2.

2.3 Ray tracing in inhomogeneous media

One approach to characterize wave propagation in inhomogeneous medium is "ray tracing". It is applicable when the wave front extends uniformly over several wavelengths and the inhomogeneity scale lengths of the medium are large in comparison to the wavelength, and particularly, when the wave has a dominant frequency. In this situation, the wave may be treated as a ray and its trajectory is tracked to explore wave propagation. The medium is then stratified into planes and the Snell's law of refraction is applied at the interface of two adjacent planes to setup the ray trajectory equation. This is illustrated by applying the arrangement shown in Fig. 2.2, in which an inhomogeneous medium with refractive index $n(z)$ is approximated by a series of plane slabs of thickness Δz, where $\Delta z \to 0$ and each slab i has a uniform refractive index n_i.

As shown in Fig. 2.2, a ray is incident from a uniform medium of refractive index n_0 at an angle, θ_0, with respect to the vertical axis into this

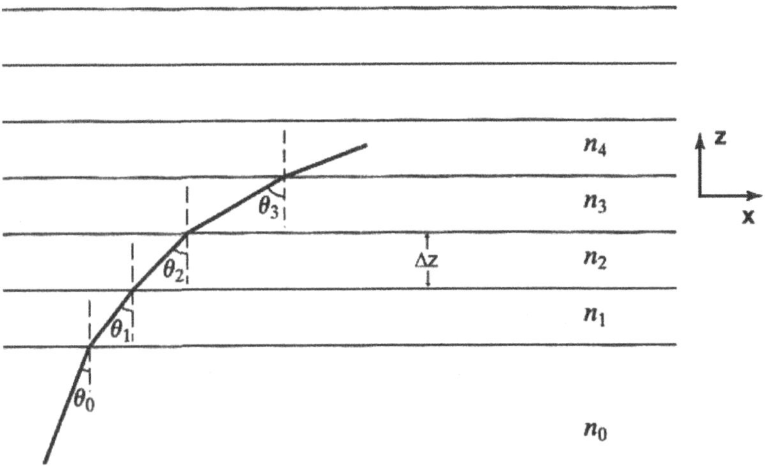

Figure 2.2. Ray trajectory in a planar stratified medium.

inhomogeneous medium; the Snell's law $n_j \sin\theta_j = n_{j+1} \sin\theta_{j+1}$ is applied at the interface of two adjacent slabs j and j+1 to determine the moving direction of the ray, $n_0 < n_1 < n_2 \ldots$ is assumed in the plot.

Because the path of the ray in each slab is a straight line, one can identify the relations: $\Delta z/\Delta x_i = \cot\theta_i$ and $n_i \sin\theta_i = n_0 \sin\theta_0$ and, where Δx_i is the horizontal displacement of the ray after transit through slab i. In the limit of $\Delta z \to 0$, these two relations become $dz/dx = \cot\theta$ and $n \sin\theta = n_0 \sin\theta_0$, and the trajectory equation is obtained to be

$$\frac{dz}{dx} = \cot\theta = \frac{(1 - \sin^2\theta)^{1/2}}{\sin\theta} = \frac{(n^2 - n_0^2 \sin^2\theta_0)^{1/2}}{n_0 \sin\theta_0}. \tag{2.8}$$

If n^2 can be modelled by a second order polynomial, i.e., $n^2(z) = a + bz + cz^2$, (2.8) can be integrated analytically; otherwise, it can be integrated numerically.

2.4 General formulation of ray trajectory equations

Consider a general case that wave propagation is governed by a dispersion equation with the generic form

$$G(\mathbf{k}, \omega;\ \mathbf{r}, t) = 0. \tag{2.9}$$

We now introduce a generic variable "τ" and take a total τ derivative on (2.9), it yields

$$\frac{dG}{d\tau} = \nabla_{k}G \cdot \frac{d\mathbf{k}}{d\tau} + \frac{\partial G}{\partial \omega}\frac{d\omega}{d\tau} + \nabla G \cdot \frac{d\mathbf{r}}{d\tau} + \frac{\partial G}{\partial t}\frac{dt}{d\tau} = 0. \tag{2.10}$$

The four terms in (2.10) are arranged into two groups to be

$$\left(\frac{\partial G}{\partial t}\frac{dt}{d\tau} + \frac{\partial G}{\partial \omega}\frac{d\omega}{d\tau}\right) + \left(\nabla G \cdot \frac{d\mathbf{r}}{d\tau} + \nabla_{k}G \cdot \frac{d\mathbf{k}}{d\tau}\right) = 0 \tag{2.11}$$

where the first group of terms is related to the time variation of the media, while the second group related to the spatial variation of the media. Because (2.11) is obtained from a general approach and the spatial variation and temporal variation are separable, its general solution requires that the following two relations be satisfied simultaneously:

$$\nabla G \cdot \frac{d\mathbf{r}}{d\tau} + \nabla_{\mathbf{k}}G \cdot \frac{d\mathbf{k}}{d\tau} = 0 \tag{2.12}$$

and

$$\frac{\partial G}{\partial t}\frac{dt}{d\tau} + \frac{\partial G}{\partial \omega}\frac{d\omega}{d\tau} = 0 \tag{2.13}$$

where $\frac{d\mathbf{r}}{d\tau} \neq \frac{d\mathbf{k}}{d\tau}$ and $\frac{dt}{d\tau} \neq \frac{d\omega}{d\tau}$ because $\mathbf{r}$ and $\mathbf{k}$ and t and ω are independent to each other, thus (2.12) and (2.13) deduce to the following relations

$$\frac{d\mathbf{r}}{d\tau} = \nabla_{\mathbf{k}}G \tag{2.14a}$$

$$\frac{d\mathbf{k}}{d\tau} = -\nabla G \tag{2.14b}$$

$$\frac{d\omega}{d\tau} = \frac{\partial G}{\partial t} \tag{2.14c}$$

$$\frac{dt}{d\tau} = -\frac{\partial G}{\partial \omega} \tag{2.14d}$$

In principle, (2.9) can be solved to obtain, for instance, $\omega = \omega(\mathbf{k}; \mathbf{r}, t)$; in other words, there are only three independent variables in (2.9). Choose $\mathbf{r}$, t, and $\mathbf{k}$ to be independent variables and ω a dependent variable, i.e., $G(\mathbf{k}, \omega; \mathbf{r}, t) = G(\mathbf{k}, \mathbf{r}, t; \omega(\mathbf{k}, \mathbf{r}, t))$ then take partial derivatives of (2.9) with respective to the three independent variables $\mathbf{k}$, $\mathbf{r}$, and t, respectively, it obtains the following three relationships

$$\nabla_{\boldsymbol{k}}\mathrm{G} + \frac{\partial \mathrm{G}}{\partial \omega}\nabla_{\boldsymbol{k}}\omega = 0 \tag{2.15a}$$

$$\nabla \mathrm{G} + \frac{\partial \mathrm{G}}{\partial \omega}\nabla \omega = 0 \tag{2.15b}$$

$$\frac{\partial \mathrm{G}}{\partial \mathrm{t}} + \frac{\partial \mathrm{G}}{\partial \omega}\frac{\partial \omega}{\partial \mathrm{t}} = 0. \tag{2.15c}$$

With the aid of (2.15a) and (2.14d), (2.14a) becomes

$$\frac{\mathrm{d}\mathbf{r}}{\mathrm{d}\tau} = \nabla_{\boldsymbol{k}}\mathrm{G} = -\frac{\partial \mathrm{G}}{\partial \omega}\nabla_{\boldsymbol{k}}\omega = \frac{\mathrm{d}\mathbf{r}}{\mathrm{dt}}\frac{\mathrm{dt}}{\mathrm{d}\tau} = -\frac{\partial \mathrm{G}}{\partial \omega}\frac{\mathrm{d}\mathbf{r}}{\mathrm{dt}},$$

which leads to

$$\frac{\mathrm{d}\mathbf{r}}{\mathrm{dt}} = \nabla_{\boldsymbol{k}}\omega = \mathbf{v}_{\mathrm{g}}. \tag{2.16}$$

With the aid of (2.15b) and (2.14d), (2.14b) becomes

$$\frac{\mathrm{d}\mathbf{k}}{\mathrm{d}\tau} = -\nabla G = \frac{\partial \mathrm{G}}{\partial \omega}\nabla \omega = \frac{\mathrm{d}\mathbf{k}}{\mathrm{dt}}\frac{\mathrm{dt}}{\mathrm{d}\tau} = -\frac{\partial \mathrm{G}}{\partial \omega}\frac{\mathrm{d}\mathbf{k}}{\mathrm{dt}},$$

which leads to

$$\frac{\mathrm{d}\mathbf{k}}{\mathrm{dt}} = -\nabla \omega. \tag{2.17}$$

With the aid of (2.15c) and (2.14d), (2.14c) becomes

$$\frac{\mathrm{d}\omega}{\mathrm{d}\tau} = \frac{\partial \mathrm{G}}{\partial \mathrm{t}} = -\frac{\partial \mathrm{G}}{\partial \omega}\frac{\partial \omega}{\partial \mathrm{t}} = \frac{\mathrm{d}\omega}{\mathrm{dt}}\frac{\mathrm{dt}}{\mathrm{d}\tau} = -\frac{\partial \mathrm{G}}{\partial \omega}\frac{\mathrm{d}\omega}{\mathrm{dt}},$$

which leads to

$$\frac{\mathrm{d}\omega}{\mathrm{dt}} = \frac{\partial \omega}{\partial \mathrm{t}}. \tag{2.18}$$

Eqs. (2.16) to (2.18), subject to a set of initial conditions, determine the ray trajectory in the phase space (i.e., $\mathbf{r} - \mathbf{k}$ space). In essence, these are the Hamilton's equations of motion with ω and $\mathbf{k}$ to be the Hamiltonian and momentum of the ray.

This ray tracing technique is illustrated in the following with two examples.

1. Consider the propagation of a sounding pulse transmitted upward from a ground based ionosonde (a HF radar); ionosonde is a remote sensing device for monitoring the plasma density profile of the bottom-side ionosphere.

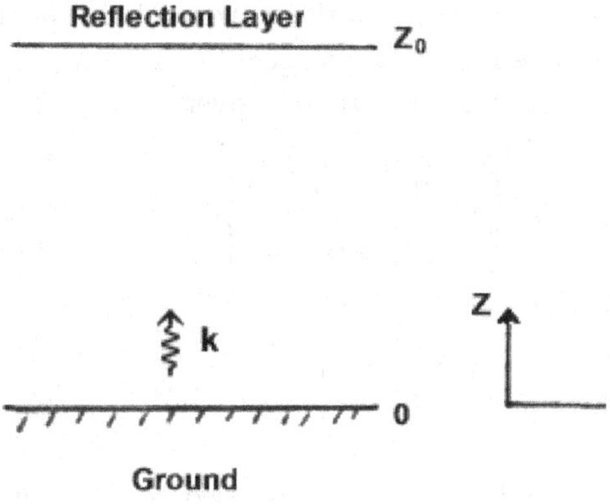

Figure 2.3. Propagation of a ray from the ground to the reflection layer at $z = z_0$.

Neglect the geomagnetic field effect, the dispersion relation is given by $\omega = (\omega_p^2 + k^2c^2)^{1/2} = \omega(z,k)$, where $\omega_p^2 = \omega_{p0}^2 z/z_0$, z_0 is the reflection height of the ray, and $\mathbf{k} = \hat{\mathbf{z}}k$ with $\hat{\mathbf{z}}$ in the upward direction as shown in Fig. 2.3. Because $\partial\omega/\partial t = 0$, (2.18) leads to $d\omega/dt = 0$; thus ω is a constant of motion and $(\omega_{p0} = \omega)$. With the aid of the explicit function of the dispersion relation, we have $\mathbf{v_g} = \mathbf{k}c^2/\omega$ and $\nabla_z\omega = \hat{\mathbf{z}}\omega_{p0}^2/2z_0\omega$, and (2.16) and (2.17) become

$$\frac{dz}{dt} = \frac{kc^2}{\omega} \tag{2.19}$$

and

$$\frac{dk}{dt} = -\frac{\omega_{p0}^2}{2z_0\omega}. \tag{2.20}$$

These two equations can be combined to a single second order differential equation

$$\frac{d^2z}{dt^2} = -\frac{c^2\omega_{p0}^2}{2z_0\omega^2} = -\frac{c^2}{2z_0}. \tag{2.21}$$

The ray trajectory is determined by solving (2.21) and (2.20), subjected to the initial conditions $(z(0) = 0)$ and $dz/dt|_{t=0} = c$, and $k(0) = k_0$ (i.e., $\omega/k_0 = c$). In upward propagation period, the trajectory is given by

$$z = z_0\left[1 - \left(1 - \frac{ct}{2z_0}\right)^2\right] \quad \text{and} \quad k = k_0\left(1 - \frac{ct}{2z_0}\right) \quad \text{for} \quad t \le t_0 \tag{2.22}$$

where $t_0 = 2z_0/c$ is the time when the wavenumber k of the ray decreases to zero and the ray is about to be reflected at $z = z_0$. For $t > t_0$, the ray propagates downward. The trajectory is given by

$$z = z_0\left\{1 - \left[\frac{c(t - t_0)}{2z_0}\right]^2\right\}$$

and (2.23)

$$k = -k_0\left[\frac{c(t - t_0)}{2z_0}\right] \quad \text{for} \quad t_0 \le t \le 2t_0.$$

The measured roundtrip time $2t_0$ is used to determine the reflection height z_0, where the plasma density is given to be $n_0 = m_e\epsilon_0\omega^2/e^2$. By sweeping the carrier frequency of the sounding pulse (e.g., from 1 to 10 MHz), the plasma density profile is obtained. It is noted that in this example the model of the ionospheric density profile is prescribed and overly simplified to achieve an analytical solution, which determines the velocity of the ray specifically. On the other hand, the ray velocity, which varies with the density profile, has to be determined self-consistently. In the measurement, the ray is assumed to propagate at free space speed "c"; hence, the height determined by the measured round trip time is called "virtual height", which can be converted to "true height" via an inverse ray tracing technique.

2. Ray Trajectory in a Graded-index (GRIN) Fiber

This example shows how a GRIN fiber guides light by trapping rays to improve the bandwidth of the fiber. Consider a graded-index fiber, its core has a parabolic profiled refractive index, $n_1 = n_0\left[1 - (x/a)^2\Delta\right]$ for $x \le a$, decreasing from the maximum value n_0 at the center, $x = 0$, to that, $n_2 = n_0(1 - \Delta)$ for $x > a$, of the cladding at the interface, $x = a$, with the cladding which has a constant refractive index n_2. This refractive index profile is illustrated in Fig. 2.4(a), which is azimuthally symmetric for a cylindrical fiber whose end-view cross section is shown in Fig. 2.4(b).

In the following, ray trajectory in a planar side-view cross section along the axis of the core, as shown in Fig. 2.4(c), is analyzed. The dispersion relation of light propagating in the core of the fiber is given to be $\omega = kc/n_1(x)$,

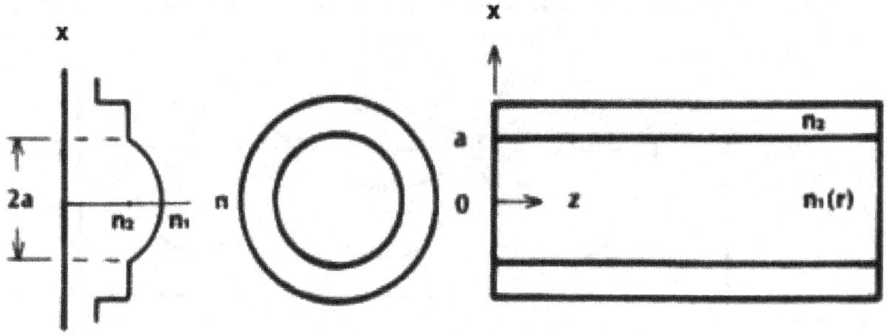

(a) (b) (c)

Figure 2.4. Graded-index fiber; (a) refractive index profile; (b) end-view and (c) side-view of the fiber cross section.

where ω is a constant of motion. Eqs. (2.16) and (2.17) become

$$\frac{dx}{dt} = \frac{k_x c}{kn_1(x)} \tag{2.24}$$

$$\frac{dz}{dt} = \frac{k_z c}{kn_1(x)} \tag{2.25}$$

$$\frac{dk_x}{dt} = -\frac{2\Delta n_0 \omega}{a^2}\frac{x}{n_1(x)} \tag{2.26}$$

$$\frac{dk_z}{dt} = 0 \tag{2.27}$$

First, (2.27) leads to $k_z = k_{z0}$; with the aid of (2.25) and (2.26), yield $dk_x/dz = (dk_x/dt)/(dz/dt) = -(2\Delta n_0 n_1 \omega^2/a^2 k_{z0} c^2)x$. Then the ratio of (2.24) to (2.25) leads to

$$\frac{d^2x}{dz^2} = \frac{1}{k_{z0}}\frac{dk_x}{dz} = -\frac{2\Delta n_0 n_1 \omega^2}{a^2 k_{z0}^2 c^2} x \tag{2.28}$$

On the right-hand side of (2.26) and (2.28), $n_1 \sim n_0$ is set because $\Delta \ll 1$. Then, (2.26) and (2.28) are solved to obtain the ray trajectory

$$x(z) = \frac{\tan\theta_0}{\beta}\sin\beta z \tag{2.29}$$

and

$$k_x(z) = k_{x0}\cos\beta z$$

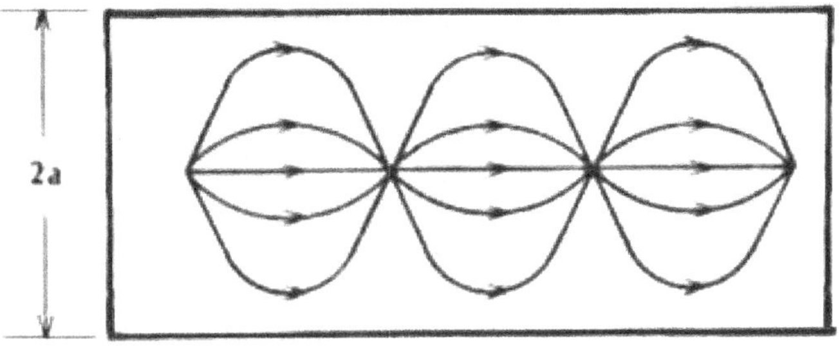

Figure 2.5. Ray trajectories along a GRIN fiber.

where θ_0 is the initial oblique angle (with respect to the axis) of the ray at $\mathrm{z} = 0$, i.e., $\tan\theta_0 = \mathrm{k_{x0}/k_{z0}}$, and $\beta = \sqrt{2\Delta}\mathrm{n_0}\omega/\mathrm{ak_{z0}c}$. When the ray is trapped in the core, $\mathrm{x(z) < a}$, θ_0 has to be smaller than a maximum angle θ_{m}, where $\theta_{\mathrm{m}} = \sin^{-1}\sqrt{2\Delta}$ is determined by the relation $\mathrm{a} = \tan\theta_{\mathrm{m}}/\beta$ from (2.29).

Figure 2.5 illustrates ray trajectories corresponding to different $|\theta_0| < \theta_m$. As shown, trapped light rays travel through the GRIN fiber in the oscillatory fashion with the same spatial period.

2.5 Method of characteristics

This is a general technique for solving first-order differential equations, which could be applied in linear, semi-linear, quasilinear, and fully nonlinear cases for not only homogeneous equations, but also inhomogeneous equations. This method converts a first-order PDE (partial differential equation) to a first-order ODE (ordinary differential equation) along a characteristic curve (or just characteristics). It is illustrated with a quasilinear PDE of the form

$$a(t)\frac{\partial}{\partial t}u + b(x)\frac{\partial}{\partial x}u = c(x,t,u). \tag{2.30}$$

First, a set of characteristic equations are defined to be

$$\frac{dt(s)}{ds} = a(t) \quad \text{and} \quad \frac{dx(s)}{ds} = b(x). \tag{2.30a}$$

Hence, along this characteristic curve (x(s), t(s)), (2.30) becomes

$$a(t)\frac{\partial}{\partial t}u + b(x)\frac{\partial}{\partial x}u = \frac{\partial u}{\partial t}\frac{dt}{ds} + \frac{\partial u}{\partial x}\frac{dx}{ds} = \frac{d}{ds}u = c(x,t,u). \tag{2.30b}$$

In the case of $a(t) = 1$, $\frac{\mathrm{dt}}{\mathrm{ds}} = 1$, which leads to $\mathrm{s} = \mathrm{t}$. A new set of characteristic equations are obtained to be

$$\frac{dx}{dt} = b(x) \quad \text{and} \quad \frac{\mathrm{d}}{\mathrm{dt}}\mathrm{u} = \mathrm{c}(\mathrm{x}, \mathrm{t}, \mathrm{u}). \tag{2.30c}$$

In other words, the original PDE in the form of

$$\frac{\partial}{\partial t}u + b(x)\frac{\partial}{\partial x}u = c(x, t, u) \tag{2.30d}$$

will be solved through its characteristics (2.30c).

Examples extended from linear homogeneous case, presented in Sec. 1.31 of Chap. 1, are illustrated in the following:

1. Linear Inhomogeneous Case

Consider single mode wave propagation in inhomogeneous media, the wave equation is given to be

$$\frac{\partial \mathrm{u}}{\partial \mathrm{t}} + \alpha \mathrm{x}^{n+1}\frac{\partial \mathrm{u}}{\partial \mathrm{x}} = 0, \tag{2.30}$$

subject to the initial condition $\mathrm{u}(\mathrm{x}, 0) = \phi(\mathrm{x})$. The characteristic equations of (2.30) are

$$\frac{\mathrm{dx}}{\mathrm{dt}} = \alpha \mathrm{x}^{n+1} \quad \text{and} \quad \frac{\mathrm{du}}{\mathrm{dt}} = 0. \tag{2.30a}$$

As indicated by the second characteristic equation in (2.30a), u(x, t) is constant along the characteristic. The trajectory of the first characteristic in (2.30a) is given to be

$$\mathrm{x}_0 = \mathrm{x}(1 + n\alpha \mathrm{x}^n t)^{-1/n},$$

where $\mathrm{x}_0 = \mathrm{x}(\mathrm{t} = 0)$ is an initial position. Thus, the solution, u(x, t), of (2.30), can be constructed, with the aid of the characteristics, via the initial condition. Explicitly,

$$\mathrm{u}(\mathrm{x}, \mathrm{t}) = \phi(\mathrm{x}_0) \quad \text{and} \quad \mathrm{x}_0 = \mathrm{x}(1 + n\alpha \mathrm{x}^n t)^{-1/n},$$

therefore, the solution of (2.30) is obtained to be

$$\mathrm{u}(\mathrm{x}, \mathrm{t}) = \phi\left[\mathrm{x}(1 + n\alpha \mathrm{x}^n t)^{-1/n}\right]. \tag{2.30b}$$

For $n \to 0$, $1 + n\alpha x^n t \to e^{n\alpha x^n t}$; hence, at $n = 0$,

$$x(1 + n\alpha x^n t)^{-1/n} = xe^{-\alpha t} = \exp(\ln x - \alpha t)$$
$$= \exp\left\{\alpha\left[\int (1/v)dx - t\right]\right\},$$

where $v = \alpha x$.

For $n \neq 0$,

$$x(1 + n\alpha x^n t)^{-1/n} = (x^{-n} + n\alpha t)^{-1/n} = \left[n\alpha(t + x^{-n}/n\alpha)\right]^{-1/n}$$
$$= \left\{n\alpha\left[t - \int (1/v)dx\right]\right\}^{-1/n},$$

where $v = \alpha x^{n+1}$. Thus, (2.30b) is a traveling wave in an inhomogeneous medium.

2. Quasi-linear Inhomogeneous Case represented by the Equation

$$\frac{\partial u}{\partial t} + \alpha x^{n+1}\frac{\partial u}{\partial x} = \beta u^2 \tag{2.31}$$

The characteristic equations of (2.31) are

$$\frac{dx}{dt} = \alpha x^{n+1} \quad \text{and} \quad \frac{du}{dt} = \beta u^2. \tag{2.31a}$$

With the aid of the solutions of the two characteristic equations:

$$x_0 = x(1 + n\alpha x^n t)^{-1/n} \quad \text{and} \quad u = u_0/(1 - \beta u_0 t),$$

where $u_0 = u(x_0, 0) = \varphi(x_0)$, the solution to (2.31) is constructed to be

$$u(x, t) = \frac{\varphi\left[x(1 + n\alpha x^n t)^{-1/n}\right]}{1 - \beta t\varphi\left[x(1 + n\alpha x^n t)^{-1/n}\right]}. \tag{2.31b}$$

As shown, (2.31b) is not a traveling wave any more. Consider a case that $n = 0$ and $\varphi(x) = \sin x$; (2.31b) becomes

$$u(x, t) = \frac{\sin(xe^{-\alpha t})}{1 - \beta t \sin(xe^{-\alpha t})}. \tag{2.31c}$$

3. Non-linear case represented by the Burgers equation

$$\frac{\partial u}{\partial t} + \alpha u\frac{\partial u}{\partial x} = 0. \tag{2.32}$$

This is the inviscid Burgers equation, which is a conservation equation; more generally, it is a first order quasilinear hyperbolic equation. The characteristic equations of (2.32) are

$$\frac{\mathrm{dx}}{\mathrm{dt}} = \alpha \mathrm{u} \quad \text{and} \quad \frac{\mathrm{du}}{\mathrm{dt}} = 0. \tag{2.32a}$$

Because u(x, t) is constant along the characteristic, i.e., $\mathrm{u(x,t)} = \varphi(\mathrm{x}_0)$, the trajectory of the first characteristic in (2.32a) is given to be $\mathrm{x}_0 = \mathrm{x} - \alpha\varphi(\mathrm{x}_0)t = \mathrm{x} - \alpha\mathrm{u(x,t)}t$. An implicit formula for the solution to (2.32) is obtained to be

$$\mathrm{u(x,\ t)} = \varphi\left[\mathrm{x} - \alpha\mathrm{u(x,t)}t\right]. \tag{2.32b}$$

It indicates that the method of characteristics is not sufficient in the nonlinear case.

2.6 Mode method for time harmonic systems

This is illustrated with a boundary value problem for a one-dimensional unforced Duffing-like equation

$$\frac{\mathrm{d}^2 q}{\mathrm{dz}^2} + \mathrm{k}^2 q - 2\alpha^2 q^3 = 0, \tag{2.33}$$

where $q = q(z)$ is subject to the boundary conditions, $q(0) = q_0$, $\dot{q}(0) = p_0$. Consider separately the cases:

a) linear: $\mathrm{k} = \text{constant}$, $\alpha = 0$
b) linear: $\mathrm{k} = \mathrm{k(z)}$, $\alpha = 0$; Eq. (2.33) reduces to (2.1)
c) nonlinear: $\mathrm{k} = \mathrm{k(z)}$, $\alpha \neq 0$

Case a) Assume the ansatz $q(z) = a\cos\theta$, where $\mathrm{d}\theta/\mathrm{dz} = \kappa$.

In other cases one can also use $q(z) = a\sin\theta$, or more generally, $q(z) = a_0 + a_1\sin\theta + a_2\cos\theta + \cdots$, the choice depends upon the boundary conditions. On substitution into (2.33), the defining equations for a and θ follow from

$$\frac{\mathrm{d}^2\mathrm{q}}{\mathrm{dz}^2} = \left(\frac{\mathrm{d}^2 a}{\mathrm{dz}^2} - a\kappa^2\right)\cos\theta - \left(2\frac{\mathrm{d}a}{\mathrm{dz}}\kappa + a\frac{\mathrm{d}\kappa}{\mathrm{dz}}\right)\sin\theta,$$

as

$$\left(\frac{\mathrm{d}^2 a}{\mathrm{dz}^2} - a\kappa^2 + a\mathrm{k}^2\right)\cos\theta - \left(2\frac{\mathrm{d}a}{\mathrm{dz}}\kappa + a\frac{\mathrm{d}\kappa}{\mathrm{dz}}\right)\sin\theta = 0, \tag{2.34}$$

whence, after setting the coefficients of the $\sin\theta$ and $\cos\theta$ terms equal to zero (since they are independent-the initial phase is arbitrary), we find

$$\kappa^2 = \mathrm{k}^2 + \frac{1}{\mathrm{a}}\frac{\mathrm{d}^2 a}{\mathrm{dz}^2} \quad \text{and} \quad \frac{\mathrm{d}}{\mathrm{dz}}(a^2\kappa) = 0. \tag{2.35}$$

Thus, (2.33) is satisfied by $a(z) = a =$ constant, $\kappa = \mathrm{k} =$ constant, and $\theta = \theta_0 + \mathrm{kz}$, where a and θ_0 are determined by the boundary conditions: $\mathrm{q}_0 = \mathrm{a}\cos\theta_0$ and $\mathrm{p}_0 = -\mathrm{k}a\sin\theta_{\mathrm{o}}$. The results are

$$\theta_0 = -\tan^{-1}\left(\frac{\mathrm{p}_0}{\mathrm{kq}_0}\right) \quad \text{and} \quad a = \sqrt{\mathrm{q}_0^2 + (\mathrm{p}_0/\mathrm{k})^2}.$$

Case b) One obtains the same defining Eqs. (2.34) for a and κ, except that $a, \kappa =$ constant is no longer a possible solution.

If k(z) is weakly variable, then so also is κ, and (hence from (2.35)) so is a. Thus, to a good approximation, one can set

$$\frac{1}{a}\frac{\mathrm{d}^2 a}{\mathrm{dz}^2} \cong 0,$$

and hence,

$$\kappa = \pm\,\mathrm{k(z)} \quad \text{and} \quad a^2(\mathrm{z})\mathrm{k(z)} = a_0^2\mathrm{k}_0, \tag{2.36a}$$

$$\theta(\mathrm{z}) = \theta_0 + \int_0^{\mathrm{z}} \mathrm{k(z')dz'}, \tag{2.36b}$$

$$a^2(\mathrm{z}) = a_0^2\frac{\mathrm{k}_0}{\mathrm{k(z)}}, \tag{2.36c}$$

where $\mathrm{k}_0 = \mathrm{k}(0)$, $a_0 = a(0) = \sqrt{q_0^2 + (\mathrm{p}_0/\mathrm{k}_0)^2}$ and $\theta_0 = -\tan^{-1}\left(\frac{p_0}{\mathrm{k}_0 q_0}\right)$.

Thus,

$$q(z) = a_0\sqrt{\mathrm{k}_0/\mathrm{k}}\cos\left(\theta_0 + \int_0^{\mathrm{z}} \mathrm{k(z')dz'}\right),$$

a WKB solution. A more accurate procedure requires an expansion in some smallness parameter.

Case c) The defining equations for a and ω become

$$\left(\frac{\mathrm{d}^2 a}{\mathrm{dz}^2} - a\kappa^2 + a\mathrm{k}^2\right)\cos\theta - \left(2\frac{\mathrm{d}a}{\mathrm{dz}}\kappa + a\frac{\mathrm{d}\kappa}{\mathrm{dz}}\right)\sin\theta$$
$$- 2\alpha^2 a^3 \cos^3\theta = 0 \tag{2.37}$$

Since

$$\cos^3\theta = \frac{3}{4}\cos\theta + \frac{1}{4}\cos 3\theta$$

then assuming that the third harmonic term $\alpha^2 a^3 \cos 3\theta$ can be neglected (because it is small and rapidly varying), and setting the independent coefficients of the $\sin\theta$ and $\cos\theta$ terms equal to zero, one finds

$$\kappa^2 = \mathrm{k}^2 - \frac{3}{2}\alpha^2 a^2 + \frac{1}{a}\frac{d^2 a}{dz^2}, \tag{2.38a}$$

and

$$\frac{d}{dz}(a^2\kappa) = 0. \tag{2.38b}$$

Again, for a weak nonlinearity (weak must be clarified in terms of a small parameter expansion), one sets $\frac{1}{a}\frac{\mathrm{d}^2 a}{\mathrm{dz}^2} \cong 0$, whence

$$\kappa = \pm\sqrt{\mathrm{k}^2 - \frac{3}{2}\alpha^2 a^2}, \tag{2.39a}$$

$$\theta = \theta_0 + \int_0^{\mathrm{z}} \kappa(\mathrm{z}')\,\mathrm{dz}', \tag{2.39b}$$

$$a^2(z) = \frac{a_0^2 \mathrm{k}_0 \sqrt{1 - \frac{3}{2}\frac{\alpha^2 a_0^2}{\mathrm{k}_0^2}}}{\mathrm{k(z)}\sqrt{1 - \frac{3}{2}\frac{\alpha^2 a^2(z)}{\mathrm{k}^2(z)}}}. \tag{2.39c}$$

where

$$a_0 = \sqrt{q_0^2 + \left(\frac{p_0}{\kappa_0}\right)^2}, \quad \theta_0 = -\tan^{-1}\left(\frac{p_0}{\kappa_0 q_0}\right), \quad \text{and} \quad \kappa_0 = \pm\sqrt{k_0^2 - \frac{3}{2}\alpha^2 a_0^2}.$$

$a(z)$ is in implicit form and (2.39c) leads to a cubic algebra equation for a^2, which is solved to explicitly obtain $a(\mathrm{z})$.

In the case of constant k, i.e., $\mathrm{k} = \mathrm{k}_0$ (2.39c) and (2.39a) leads to $a(z) = a_0$ and $\kappa_0 = \left(k^2 - \frac{3}{2}\alpha^2 a_0^2\right)^{1/2}$. Thus, the approximate solution of (2.33) is obtained to be

$$q(z) = a_0 \cos(\theta_0 + \kappa_0 \mathrm{z}) \tag{2.39d}$$

which is valid under the assumption that $(\alpha a_0/\mathrm{k})^2 \ll 1$.

It is of interest to observe, upon multiplication of (2.33) by $\dot{q}$, that

$$\frac{d}{dz}\left(\frac{\dot{q}^2}{2} + \frac{\mathrm{k}^2 q^2}{2} - \frac{\alpha^2 q^4}{2}\right) = 0, \quad \text{for constant k}. \tag{2.40}$$

Therefore, the value $1/2\ (\dot{q}^2 + \mathrm{k}^2 q^2 - \alpha^2 q^4) = H$ of the bracketed expression, representing the energy density of a wave, is a constant. This is because the system is lossless and uniform. The approximate solution (2.39d), which neglects the harmonic terms, cannot fulfill the requirement.

In the homogeneous situation, i.e., k and α are constants, and when the wave amplitude is limited in the stable regime, i.e., $H \le \alpha\frac{\mathrm{k}^4}{8\alpha^2}$, Eq. (2.33) has analytical mode solutions.

1) In the case of $H < \frac{\mathrm{k}^4}{8\alpha^2} = \frac{k^2 q_M^2}{4}$, a periodic mode solution is given to be

$$q(z) = q_2 \,\mathrm{sn}\left(\frac{q_1}{\sqrt{2}q_M}\mathrm{k}(\mathrm{z}+\mathrm{z}_0), \beta^2\right), \tag{2.41}$$

where "sn" is a Jacobi elliptic function, $q_M = \frac{\mathrm{k}}{\sqrt{2}\alpha}$,

$$q_1 = q_M\left(1 + \sqrt{1 - \frac{4H}{k^2 q_M^2}}\right)^{\frac{1}{2}},$$

$$q_2 = q_M\left(1 - \sqrt{1 - \frac{4H}{k^2 q_M^2}}\right)^{\frac{1}{2}},$$

and $\beta = \frac{q_2}{q_1}$. z_0 is determined by the boundary condition:

$$q_2 \,\mathrm{sn}\left(\frac{q_1}{\sqrt{2}q_M}\mathrm{k}z_0, \beta^2\right) = q_0.$$

In the case of $\beta \cong \alpha\sqrt{2H}/\mathrm{k}^2 \ll 1$; (2.41) reduces to (2.39d).

2) When $H = \frac{\mathrm{k}^4}{8\alpha^2} = \frac{k^2 q_M^2}{4}$, $q_2 = q_M = q_1$, and $\beta = 1$; hence,

$$q(z) = q_M \tanh\frac{\mathrm{k}}{\sqrt{2}}(\mathrm{z}+\mathrm{z}_0). \tag{2.42}$$

This is an aperiodic mode and z_0 is determined by the boundary condition:

$$q_0 = \frac{\mathrm{k}}{\sqrt{2}\alpha}\tanh\frac{\mathrm{k}}{\sqrt{2}}\mathrm{z}_0.$$

Whitham's average Lagrangian and Hamiltonian method of describing the problem will be introduced in Chap. 5 to rephrase the similar equation being considered.

Problems

P2.1. Consider the propagation of a plane wave $E(z \to -\infty) = E_0 e^{ik_0 z}$ transmitted from the ground upward vertically to the ionosphere, assuming that the ground is located near $z \to -\infty$. The plasma density is modelled by an Epstein density profile $\mathrm{n(z) = n_m sech^2}\left(\frac{z}{L}\right)$, where $n_m = n(z = 0)$ is the peak plasma density at foF2 layer of the ionosphere; the inhomogeneity scale length $\mathrm{L} >> \lambda_0 = \frac{2\pi}{k_0}$; $k_0 = \frac{\omega_0}{c}$ and the wave frequency $\omega_0 = \omega_{pm}$ is operated. Thus, $k^2(z) = k_0^2 \tanh^2(z/L)$. Find E(z) in the ionosphere.

P2.2. Derive explicitly the ray tracing equations for an EM wave propagating in inhomogeneous unmagnetized plasma; the dispersion relation of the wave in plasma is given to be

$$\omega = \sqrt{\omega_p^2(z) + k^2c^2}.$$

P2.3. Ray trajectory in the bottom-side of the ionosphere: in the case of oblique incidence with the wavevector $\mathbf{k} = \hat{\mathbf{x}}k_x + \hat{\mathbf{z}}k_z$, where $\hat{\mathbf{z}}$ is in the upward direction and $\hat{\mathbf{x}}$ in the horizontal direction, with the initial condition $\mathbf{k}(0) = \hat{\mathbf{x}}k_{x0} + \hat{\mathbf{z}}k_{z0} = \mathbf{k}_0$; $k_0 = \sqrt{k_{x0}^2 + k_{z0}^2} = \omega/c$. The dispersion relation $\omega = \sqrt{\omega_p^2(z) + k_x^2c^2 + k_z^2c^2} = \omega(z, \mathbf{k})$, where $\omega_p^2(z) = \omega_{p0}^2\, z/z_0$, z_0 is the reflection height of the ray, and $\omega_{p0}^2 = \omega^2 - k_{x0}^2c^2$; the ray starts at $x(0) = 0$ and $z(0) = 0$.

(1) Show that the trajectory is given by $z/z_0 = 1 - \left(1 - \frac{1}{2}\alpha x/z_0\right)^2$, where $\alpha = \omega_{p0}/k_{x0}c$;
(2) plot the trajectory z/z_0 vs x/z_0, i.e., on the x-z plane, of one round trip for incident angle of 30°.

P2.4. Consider a plasma lens having a parabolic density distribution, $n_e = n_0(y/y_0)^2$, in y, it guides a ray to propagate along its central (symmetrical) axis in the x direction as shown in Fig. P2.1. The dispersion relation is given to be $\omega = \left[\omega_p^2(y) + k_x^2c^2 + k_y^2c^2\right]^{1/2} = \omega(y, k_x, k_y)$, where $\omega_p^2 = \omega_{p0}^2 y^2/y_0^2$ and $\omega_{p0} = (n_0 e^2/m_e\varepsilon_0)^{1/2}$. In the case with the initial conditions $x(0) = x_0$, $y(0) = 0$, and $\mathbf{k}(0) = \hat{\mathbf{x}}k_{x0} + \hat{\mathbf{y}}k_{yo}$, where $k_0 = (k_{x0}^2 + k_{y0}^2)^{1/2} = \omega/c$. Find the ray trajectory in the plasma lens.

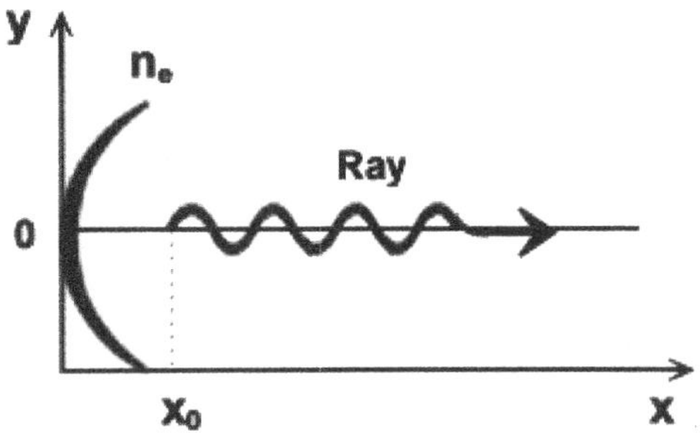

Figure P2.1. Propagation of a ray along the central axis of a plasma lens that has a parabolic density distribution in the transverse (y) direction.

P2.5. Show that the ray trajectory in Fig. P2.1 bounces back and forth in y direction at a bounce frequency $\omega_B = (\omega_{p0}c/\omega y_0)$ i.e., $y(t) = y_0 \sin\omega_B t$.

P2.6. The plasma density distribution in the bottom-side of the ionosphere is modelled by an Epstein profile to be $n(z) = n_0\left(1 - e^{-\frac{z}{L}}\right)$, where $z = 0$ is located at the lower boundary of the ionosphere. Plasma has a dispersion $\omega = \sqrt{\omega_p^2 + k^2c^2}$, where $\omega_p^2 = \omega_0^2\left(1 - e^{-\frac{z}{L}}\right)$ and $\omega_0^2 = \frac{n_0 e^2}{m_e \epsilon_0}$. A wave at frequency ω_0 is transmitted upward vertically from the ground into the ionosphere, find the ray trajectory $(z(t), \mathbf{k}(t))$ in the ionosphere.

P2.7. Apply the characteristic equations (2.30a) to solve (2.30) for $a(t) = t, b(x) = -2x$, and $c(x, t, u) = u^2$ and the initial condition $u(x, t = 1) = x$.

P2.8. Find the solution of Eq. (2.31), in the case of $n = 1$, for the initial condition $u_0 = \cos kx$.

P2.9. The inviscid Burgers equation (2.32) has the initial condition $u(x, 0) = ax^2 + bx + c$, find the solution $u(x, t)$

Chapter 3

Waves Traversing a Temporal Discontinuity Interface Between Two Media

3.1 Space-time duality of wave phenomena at a discontinuity interface between media

When electromagnetic (EM) wave propagation encounters a discontinuity of the medium, backward and forward propagating waves at the boundary interface are generated to manifest the reflection and transmission phenomena. In the case of spatial boundary, such as when a wave propagates from one dielectric medium to another dielectric medium, the wave frequency does not change; the wavelength of the transmitted wave is changed, but the wavelength of the reflected wave remains the same. The time domain discontinuity (temporal boundary) may be established by suddenly creating a dielectric medium (e.g., plasma) in the background. In this case, it is the wavelength that does not change. Moreover, a temporal interface is a boundary across which the refractive index changes for the entire medium, thus the frequencies of both the transmitted (forward propagating) and reflected (backward propagating) waves are changed because the causality forbids the reflected wave to access the initial medium.

It is difficult to observe time domain wave transition when propagating in a solid dielectric medium because it is not likely to change its dielectric property suddenly. On the other hand, gaseous plasma is a dielectric medium and can be produced rapidly over a large volume. In the following, time-domain wave phenomena are illustrated by considering wave propagation in suddenly created plasmas.

3.2 Wave propagation in suddenly created unmagnetized plasma

Propagation of electromagnetic (EM) wave in unmagnetized uniform plasma is governed by the wave equation

$$\left[\partial_z^2 - \frac{1}{c^2}\partial_t^2 - \frac{\omega_p^2}{c^2}\right]\varepsilon(z,t) = 0 \tag{3.1}$$

where $\partial_t = \frac{\partial}{\partial t}$ and $\partial_z = \frac{\partial}{\partial z}$; $\omega_p = \sqrt{\frac{N_0 e^2}{m_e \epsilon_0}}$ is the electron plasma frequency, N_0 is the electron density, e is the magnitude of the electron charge, m_e is the electron mass, and ϵ_0 is the permittivity of the free space; $\mathcal{E}(z,t)$ is the wave electric field.

In this linear system, the basic source free wave (mode) types can be characterized by the space-time harmonic solution of the form $\exp[i(kz - \omega t)]$, which is substituted into (3.1) to obtain the dispersion relation

$$\omega(k) = \sqrt{\omega_p^2 + k^2c^2}. \tag{3.1a}$$

As shown in Fig. 3.1, the dispersion curve $\omega(k)$ in the ω-k diagram is a parabola in the presence of plasma and reduces to a straight line

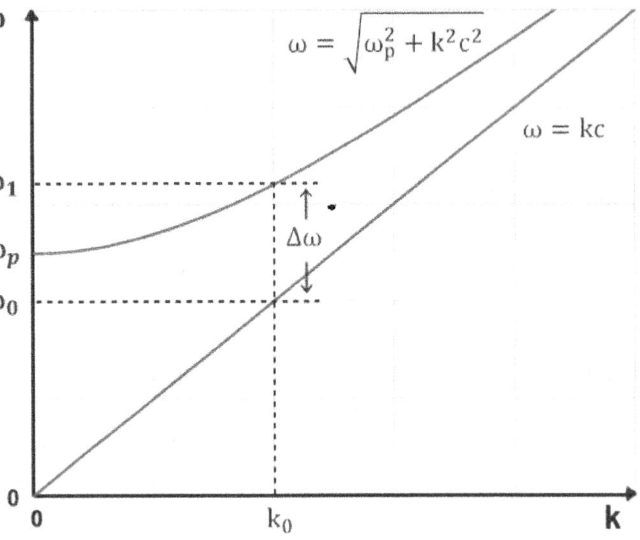

Figure 3.1. Dispersion relations of EM wave propagating in free space and in plasma.

$\omega = \mathrm{kc}$ in the absence of plasma. If plasma is suddenly created at $\mathrm{t} = 0$, a pre-existing plane wave of the form: $\exp[i(\mathrm{k_0 z} - \omega_0 \mathrm{t})]$, where $\omega_0 = \mathrm{k_0 c}$, in $\mathrm{t} < 0$, has to shift the frequency to

$$\omega = \sqrt{\omega_p^2 + k_0^2 c^2} \tag{3.1b}$$

in order to satisfy the new dispersion relation in plasma defined by the parabola. The wavelength λ_0 of the wave remains invariant during the temporal discontinuity.

A case that a wave exists everywhere in free space prior to the plasma creation at $\mathrm{t} = 0$ is considered; the wave fields are

$$\begin{cases} \boldsymbol{E}(\mathrm{z}, \mathrm{t} \le 0) = \hat{\mathbf{x}} E_0 \cos(\mathrm{k_0 z} - \omega_0 \mathrm{t}) \\ \boldsymbol{H}(\mathrm{z}, \mathrm{t} \le 0) = \hat{\mathbf{y}} \left(\dfrac{E_0}{\eta_0}\right) \cos(\mathrm{k_0 z} - \omega_0 \mathrm{t}) \end{cases} \tag{3.2}$$

where $\mathrm{k_0} = \omega_0/\mathrm{c}$. The plasma is modelled by

$$\begin{cases} \omega_p^2(z,t) = 0 & \text{in } t < 0 \\ \omega_p^2(z,t) = \omega_p^2 = \text{const.} & \text{in } t > 0 \end{cases} \tag{3.3}$$

which is suddenly created at $\mathrm{t} = 0$. In this case, Eq. (3.1) is solved analytically.

Propagation in the new medium in $\mathrm{t} > 0$, waves are governed by the new dispersion relation (3.1a). Because the plasma is created instantaneously and uniformly, the wavelength of the original wave is not affected, i.e., $\mathbf{k} = \mathbf{k}_0$, and the spatial distributions of the wave fields $E(\mathrm{z}, \mathrm{t} = 0^+)$ and $H(\mathrm{z}, \mathrm{t} = 0^+)$ at $t = 0^+$ will be the same as the initial distributions $E(\mathrm{z}, \mathrm{t} = 0^-) = E_0 \cos \mathrm{k_0 z}$ and $H(\mathrm{z}, \mathrm{t} = 0^-) = (E_0/\eta_0) \cos \mathrm{k_0 z}$ *at* $t = 0^-$.

The original wave is converted to both forward propagating and backward propagating waves at the same upshift-frequency (3.1b). The wave electric field is expressed to be

$$\boldsymbol{E}(\mathrm{z}, \mathrm{t} > 0) = \hat{\mathbf{x}} \left[A_+ \cos(\mathrm{k_0 z} - \omega \mathrm{t}) + A_- \cos(\mathrm{k_0 z} + \omega \mathrm{t})\right] \tag{3.4}$$

where

$$A_+ + A_- = E_0 \tag{3.5}$$

is given by the initial condition of wave electric field.

From the Faraday's Law

$$\nabla \times \mathbf{E} = -\mu_0\, \partial_t \mathbf{H},$$

the corresponding wave magnetic field $\boldsymbol{H}$ is obtained to be

$$\boldsymbol{H}(\mathrm{z},\, \mathrm{t} > 0) = \hat{\mathbf{y}} \left(\frac{\mathrm{k}_0}{\mu_0 \omega} \right) [A_+ \cos(\mathrm{k}_0 \mathrm{z} - \omega \mathrm{t}) - \mathrm{A}_- \cos(\mathrm{k}_0 \mathrm{z} + \omega \mathrm{t})] + \boldsymbol{H}_{\mathrm{w}}(\mathrm{z}) \tag{3.6}$$

where $\boldsymbol{H}_{\mathrm{w}}$ (z) is a time integration constant. A space charge current density, $\boldsymbol{J} = -\mathrm{eN}_0\mathbf{v}_\mathrm{e}$, associated with the fields (3.4) and (3.6) will be induced in plasma. Their relationship is governed by the Ampere's law,

$$\mathbf{J} = -\epsilon_0 \partial_\mathrm{t} \mathbf{E} + \nabla \times \mathbf{H}.$$

With the aid of (3.4) and (3.6), this induced current density is obtained to be

$$\boldsymbol{J}(z,t) = -\hat{\mathbf{x}} \left(\frac{\epsilon_0 \omega_p^2}{\omega} \right) [A_+ \sin(\mathrm{k}_0 \mathrm{z} - \omega_0 \mathrm{t}) - A_- \sin(\mathrm{k}_0 \mathrm{z} + \omega_0 \mathrm{t})] + \nabla \times \boldsymbol{H}_{\mathrm{w}}(\mathrm{z}) \tag{3.7}$$

Although plasma is conducting, it takes time for plasma to response to the wave fields. At $t = 0^+$, plasma is just created and has no time to response to the wave fields, thus no current is generated at $t = 0^+$, i.e., $\boldsymbol{J}(z, t = 0^+) = 0$, and (3.7) reduces to

$$\nabla \times \boldsymbol{H}_w(\mathrm{z}) = \hat{\mathbf{x}} \left(\frac{\epsilon_0 \omega_p^2}{\omega} \right) (A_+ - A_-) \sin k_0 \mathrm{z}. \tag{3.8}$$

Eq. (3.8) is solved to obtain

$$\boldsymbol{H}_w(z) = \hat{\mathbf{y}} \frac{\epsilon_0 \omega_p^2}{k_0 \omega} (A_+ - A_-) \cos k_0 \mathrm{z}. \tag{3.9}$$

The initial condition $H(\mathrm{z}, \mathrm{t} = 0^+) = (E_0/\eta_0) \cos \mathrm{k}_0 \mathrm{z}$ is then applied to (3.6), with the aid of (3.9), to obtain the relation

$$A_+ - A_- = \left(\frac{\omega_0}{\omega} \right) E_0. \tag{3.10}$$

The initial conditions (3.5) and (3.10) comply with the continuity of the power flow and the conservation of energy at $\mathrm{t} = 0$; in other words, the

Poynting vector $\mathbf{S} = \boldsymbol{E} \times \boldsymbol{H}$ and the field energy density $\varepsilon = {}^1\!/_2\epsilon_0 E^2 + {}^1\!/_2\,\mu_0 H^2$, at $t = 0^-$ and at $t = 0^+$ are equal:

$$\boldsymbol{S}(\mathrm{z}, t = 0^-) = \boldsymbol{S}(\mathrm{z}, t = 0^+) = \hat{\mathbf{z}}\left(\frac{E_0^2}{\eta_0}\right)\cos^2 k_0 z,$$

and

$$\varepsilon(\mathrm{z}, t = 0^-) = \varepsilon(\mathrm{z}, t = 0^+) = \epsilon_0 E_0^2 \cos^2 k_0 z;$$

it is noted that at $t = 0$, no field energy is stored in the just created plasma.

Eqs. (3.5) and (3.10) are solved to obtain

$$A_+ = (1 + \omega_0/\omega)E_0/2 \quad \text{and} \quad A_- = (1 - \omega_0/\omega)E_0/2,$$

and time domain reflection coefficient Γ_t and transmission coefficient $\mathrm{T_t}$ are obtained to be $\Gamma_\mathrm{t} = A_-/E_0 = (1 - \omega_0/\omega)/2$ and $\mathrm{T_t} = A_+/E_0 = (1 + \omega_0/\omega)/2$, respectively. It shows that $\Gamma_\mathrm{t} + \mathrm{T_t} = 1$, satisfying the continuity condition of the wave field at the temporal discontinuity; it is different from the spatial domain continuity condition $\mathrm{T} = 1 + \Gamma$ of the wave field at the interface of the spatial discontinuity.

But the wave transition also complies with the conservation of momentum. The momentum density $\mathbf{P}_\mathrm{v}$ of the wave is given by

$$\mathbf{P}_\mathrm{v} = \mathbf{S}_\mathrm{v}/\mathrm{v}_\mathrm{g}^2,$$

where $\mathbf{S}_\mathrm{v} = \mathbf{E}_\mathrm{v} \times \mathbf{H}_\mathrm{v}$ is the Poynting vector of the wave and $\mathrm{v_g} = \partial\omega/\partial k$ is the group velocity.

At $t = 0^-$, $\mathrm{v_g} = \mathrm{c}$ and $\mathbf{S}_{\mathrm{v}<} = \hat{\mathbf{z}}(E_0^2/\eta_0)\cos^2 k_0 z$, giving

$$\mathbf{P}_{\mathrm{v}<} = \hat{\mathbf{z}}(E_0^2/\eta_0 c^2)\cos^2 k_0 z;$$

at $t = 0^+$, $\mathrm{v_g} = (\omega_0/\omega)\mathrm{c}$ and $\mathbf{S}_{\mathrm{v}>} = \hat{\mathbf{z}}(\mathrm{k}_0/\mu_0\omega)(\mathrm{A}_+^2 - \mathrm{A}_-^2)\cos^2 k_0 z$, thus

$$\begin{aligned}\mathbf{P}_{\mathrm{v}>} &= \hat{\mathbf{z}}\frac{\omega}{\omega_0\eta_0 c^2}(\mathrm{A}_+^2 - \mathrm{A}_-^2)\cos^2 \mathrm{k}_0 z \\ &= \hat{\mathbf{z}}\frac{E_0^2}{\eta_0 c^2}\cos^2 k_0 z = \mathbf{P}_{\mathrm{v}<}\,.\end{aligned}$$

In the transition of wave conversion, some of the original wave energy is converted to the energy of the wiggler magnetic field (3.7), with the conversion ratio $\zeta_\mathrm{w} = {}^1\!/_2(\omega_\mathrm{p}/\omega)^4$. The appearance of an additional wiggler magnetic field is a distinct wave phenomenon in the time-domain transition.

Figure 3.1 shows that there is only one EM branch of modes in unmagnetized plasma; as a magnetic field is embedded in the background, the suddenly created plasma may support multiple branches of EM modes; additional distinctive in the time domain wave transition is illustrated in the following example.

3.3 Wave propagation in suddenly created magneto plasma

Assume that ions are immobile, the relation, $\mathbf{J} = -\mathrm{eN_0}\mathbf{v_e} = \epsilon_0\omega_\mathrm{p}^2(\partial_\mathrm{t} - \Omega_\mathrm{e}\hat{\mathbf{z}}\times)^{-1}\boldsymbol{E}$, deduced from the electron momentum equation, is substituted into the Maxwell's equations. The wave equation for the propagation of electromagnetic wave along the magnetic field $\hat{\mathbf{z}}\mathrm{B_0}$ in uniform plasma is modified from (3.1) to

$$\partial_\mathrm{t}\left(\partial_\mathrm{z}^2 - \frac{1}{\mathrm{c}^2}\partial_\mathrm{t}^2 - \frac{\omega_\mathrm{p}^2}{\mathrm{c}^2}\right)\boldsymbol{E}(\mathrm{z},\mathrm{t}) + \Omega_\mathrm{e}\left(\partial_\mathrm{z}^2 - \frac{1}{\mathrm{c}^2}\partial_\mathrm{t}^2\right)\boldsymbol{E}(\mathrm{z},\mathrm{t}) \times \hat{\mathbf{z}} = 0 \quad (3.11)$$

where $\Omega_\mathrm{e} = \mathrm{eB_0/m_e}$ is the electron cyclotron frequency, $\mathrm{m_e}$ is the electron mass, and $\boldsymbol{E}(\mathrm{z}, \mathrm{t})$ is the wave electric field.

3.3.1 *Branches of modes*

In the absence of plasma, i.e., $\omega_\mathrm{p} = 0$, (3.11) reduces to the free space wave equation $\left(\partial_\mathrm{z}^2 - \frac{1}{\mathrm{c}^2}\partial_\mathrm{t}^2\right)\boldsymbol{E}(\mathrm{z},\mathrm{t}) = 0$. A forward/backward propagating mode $\varepsilon^{\pm}(\mathrm{z},\mathrm{t}) = \mathrm{Re}\{\boldsymbol{A}\exp[\mathrm{i}(\mathrm{kz} \mp \omega\mathrm{t})]\}$, where the superscripts "$\pm$", representing forward/backward propagations, correspond to the "$\mp$" signs in the right hand side (RHS) parenthesis; "Re" represents the real part in the parenthesis, and $\boldsymbol{A}$ is a constant vector on the x-y plane, is governed by the dispersion relation $\omega = \mathrm{kc}$. In unmagnetized plasma, i.e., $\omega_\mathrm{p} \neq 0$ and $\mathrm{B_0} = 0$, (3.11) reduces to (3.1), and the dispersion relation becomes (3.1a), which is plotted in Fig. 3.1.

For right/left-hand circularly polarized plane waves of $\boldsymbol{E}_\pm(\mathrm{z},\mathrm{t}) = \mathrm{Re}\{[(\hat{\mathbf{x}} \pm \mathrm{i}\hat{\mathbf{y}})E\exp[i(kz - \omega t)]\}$ propagating forward along the magnetic field, where the subscripts "$\pm$" signs represent right/left-hand circular polarization (RHCP/LHCP), corresponding to the "$\pm$" signs in the RHS parenthesis, the frequency ω and the wavenumber k are governed by the dispersion relations of the R-mode/L-mode

$$k(\omega) = (\omega/\mathrm{c})\sqrt{1 - \omega_\mathrm{p}^2/\omega(\omega \mp \Omega_\mathrm{e})} \quad (3.12)$$

where "$\mp$" signs correspond to the R-mode/L-mode, respectively. It is noted that if the propagation direction is opposite to the background magnetic field, RHCP/LHCP waves become L-mode/R-mode.

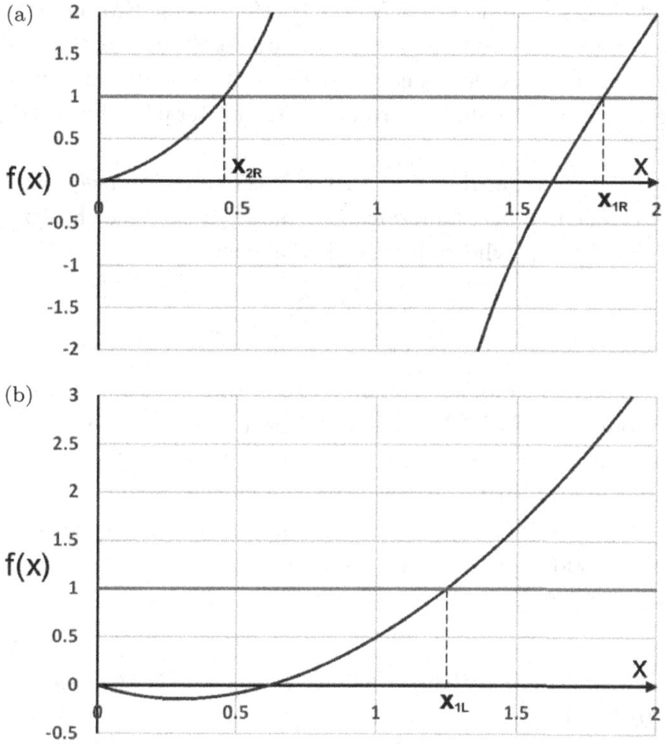

Figure 3.2. A graphical means to solve the dispersion equations; (a) frequencies ($\omega_{1R} \propto X_{1R}$ & $\omega_{2R} \propto X_{2R}$) of the R-mode and slow R-mode at a given wavelength and (b) frequency ($\omega_{1L} \propto X_{1L}$) of the L-mode at the same wavelength as that of (a).

Introduce the dimensionless variable and parameters: $x = \omega/kc, \alpha = \omega_p^2/k^2c^2$, and $\beta = \Omega_e/kc$, (3.12) reduces to

$$1 = x^2\left[1 - \frac{\alpha}{x(x \mp \beta)}\right] = f(x) \tag{3.13}$$

which is solved graphically as illustrated in the following.

Consider a case of $\alpha = \beta = 1$, the solutions of (3.13) are identified by the intersection points shown in Figs. 3.2(a) and (b) for RHCP/LHCP waves, respectively. In Fig. 3.2(a), there are two intersection points corresponding to two different R-modes, one intersection point at $x = x_{1R} = 1.802$ is a fast (R-mode) EM mode in the branch of $x \geq x_{cr} = 1.618$, i.e., $\omega_{1R} \geq \omega_{cr}$, where the cutoff frequency

$$\omega_{cr} = \Omega_e/2 + \sqrt{\omega_p^2 + \Omega_e^2/4};$$

another intersection point at $x = x_{2R} = 0.445$ is a slow EM mode (i.e., slow R-mode) in the same dispersion branch of a whistler mode, $0 < x < 1$ (i.e., $0 < \omega_{2R} < \Omega_e$). The emergence of this slow R-mode branch in magneto plasma is attributed to the electron cyclotron interaction with the RHCP wave.

In Fig. 3.2(b) regarding the LHCP case, one intersection point at $x = x_{1L} = 1.247$ is a fast (L-mode) EM mode in the branch of $x \geq x_{c\ell} = 0.618$, i.e., $\omega_{1L} \geq \omega_{c\ell}$, where the cutoff frequency

$$\omega_{c\ell} = -\Omega_e/2 + \sqrt{\omega_p^2 + \Omega_e^2/4}.$$

There is no slow L-mode branch because the gyration of electrons is reversely from that of LHCP wave and ions are assumed to be immobile.

It is noted that the fast EM (R-mode and L-mode) waves can propagate in free space; on the other hand the slow R-mode (e.g., a whistler mode) is supported by the plasma, it doesn't exist in the absence of plasma and magnetic field. The mode frequency has to be lower than the electron cyclotron frequency Ω_e.

3.3.2 *Continuity conditions at temporal discontinuity interface*

The transition of RHCP/LHCP plane waves, propagating along the background magnetic field $\hat{\mathbf{z}}B_0$, through a suddenly created uniform plasma, is analyzed in the following. Before the plasma creation at $t = 0$, these waves are propagating in the free space (i.e., $\omega_p^2(t < 0) = 0$) with the wave fields

$$\begin{aligned}\mathbf{E}_\pm(z, t \leq 0) &= E_0[\hat{\mathbf{x}}\cos(k_0 z - \omega_0 t) \mp \hat{\mathbf{y}}\sin(k_0 z - \omega_0 t)] \\ &= \text{Re}\,\{(\hat{\mathbf{x}} \pm i\hat{\mathbf{y}})E_0 \exp[i(k_0 z - \omega_0 t)]\} \end{aligned} \tag{3.14a}$$

and

$$\begin{aligned}\mathbf{H}_\pm(z, t \leq 0) &= \pm\left(\frac{E_0}{\eta_0}\right)[\hat{\mathbf{x}}\sin(k_0 z - \omega_0 t) \pm \hat{\mathbf{y}}\cos(k_0 z - \omega_0 z)] \\ &= \text{Re}\left\{\mp i(\hat{\mathbf{x}} \pm i\hat{\mathbf{y}})\left(\frac{E_0}{\eta_0}\right)\exp[i(k_0 z - \omega_0 t)]\right\} \end{aligned} \tag{3.14b}$$

where $k_0 = \omega_0/c$ is the wavenumber and $\eta_0 = \sqrt{\mu_0/\epsilon_0}$ is the intrinsic impedance of the free space. Substitute (3.14a) into (3.11), yields

$$\mathrm{Re}\left\{(\hat{\mathbf{x}} \pm i\hat{\mathbf{y}})(\omega_0^2 \mp \omega_0\Omega_e - \Omega_e\Omega_i)\left(k_0^2 - \frac{\omega_0^2}{c^2}\right)E_0 \exp\left[i(k_0 z - \omega_0 t)\right]\right\} = 0$$

which conforms that (3.14) satisfies (3.11).

At $t = 0$, a uniform plasma is suddenly created in the background with $\omega_p^2(t \geq 0) = \omega_p^2 = \text{const.}$; it will not affect the wavelength of the original wave, and the spatial distributions of the wave fields $\mathbf{E}_\pm(z, t)$ and $\mathbf{H}_\pm(z, t)$ at $t = 0^+$ will stay the same as the distributions $\mathbf{E}_\pm(z, t = 0^-)$ and $\mathbf{H}_\pm(z, t = 0^-)$ at $t = 0^-$, i.e.,

$$\begin{aligned}\mathbf{E}_\pm(z, t = 0^+) &= \mathbf{E}_\pm(z, t = 0^-)\\ &= E_0[\hat{x}\cos k_0 z \mp \hat{y}\sin k_0 z]\end{aligned} \tag{3.15a}$$

and

$$\begin{aligned}\mathbf{H}_\pm(z, t = 0^+) &= \mathbf{H}_\pm(z, t = 0^-)\\ &= \left(\frac{E_0}{\eta_0}\right)[\hat{y}\cos k_0 z \pm \hat{x}\sin k_0 z]\end{aligned} \tag{3.15b}$$

but, in $t > 0$, the wave field (3.14) does not satisfy (3.11) anymore.

The continuity condition (3.15a) imposes the polarizations of the converted waves; and the dispersion relations (3.12) with $k = k_0 = \omega_0/c$ determine the frequencies of converted waves. Thus, in each polarization case, the converted forward and backward waves have the same polarization (i.e., different mode types), e.g., a RHCP/LHCP original wave will convert to RHCP/LHCP forward waves and RHCP/LHCP backward waves.

(1) In the RHCP case, the original wave will convert to two forward (R-mode and slow R-mode at frequencies ω_{1R} and ω_{2R}) and one backward (L-mode at frequency ω_{1L}) propagating waves which satisfy (3.11).
(2) In the LHCP case, the original wave will convert to one forward (L-mode at frequency ω_{1L}) and two backward (R-mode and slow R-mode at frequencies ω_{1R} and ω_{2R}) propagating waves which satisfy (3.11).

Superscripts "±", subscripts "±", and subscripts "R, L", are used in the following to distinct the forward/backward propagation directions,

RHCP/LHCP, and R-/L-mode types, respectively. It is noted again that in the forward propagation (indicated by superscript "+"), i.e., along the background magnetic field, RHCP/LHCP waves become R-mode/L-mode; on the other hand, in the backward propagation (indicated by superscript "−"), i.e., opposite to the background magnetic field, RHCP/LHCP waves become L-mode/R-mode.

Therefore, in t > 0, the wave electric field is expressed to be

$$\begin{aligned}\mathbf{E}_{\pm}(\mathrm{z},\,\mathrm{t}>0) &= \sum_{\mathrm{i}=1}^{2}\{E_{iR,L}^{+}[\hat{\mathbf{x}}\cos(\mathrm{k_0z}-\omega_{\mathrm{iR,L}}\mathrm{t}) \mp \hat{\mathbf{y}}\sin(\mathrm{k_0z}-\omega_{\mathrm{iR,L}}\mathrm{t})] \\ &\quad + E_{iL,R}^{-}[\hat{\mathbf{x}}\cos(\mathrm{k_0z}+\omega_{\mathrm{iL,R}}\mathrm{t}) \mp \hat{\mathbf{y}}\sin(\mathrm{k_0z}+\omega_{\mathrm{iL,R}}\mathrm{t})]\}\end{aligned} \tag{3.16}$$

where $E_{2L}^{+} = E_{2L}^{-} = 0$, because there is no slow L-mode. Substituting (3.16) into the Maxwell–Faraday equation $\nabla \times \mathbf{E} = -\mu_0\partial_\mathrm{t}\mathbf{H}$, it is solved to obtain the magnetic field $\mathbf{H}$ to be

$$\begin{aligned}\mathbf{H}_{\pm}(\mathrm{z},\,\mathrm{t}>0) &= \sum_{\mathrm{i}=1}^{2}\left\{\left(\frac{E_{iR,L}^{+}}{\eta_{\mathrm{iR,L}}}\right)[\hat{\mathbf{y}}\cos(\mathrm{k_0z}-\omega_{\mathrm{iR,L}}\mathrm{t})] \pm \hat{\mathbf{x}}\sin(\mathrm{k_0z}-\omega_{\mathrm{iR,L}}\mathrm{t})]\right. \\ &\quad \left. -\left(\frac{E_{iL,R}^{-}}{\eta_{\mathrm{iL,R}}}\right)[\hat{\mathbf{y}}\cos(\mathrm{k_0z}+\omega_{\mathrm{iL,R}}\mathrm{t}) \pm \hat{\mathbf{x}}\sin(\mathrm{k_0z}+\omega_{\mathrm{iL,R}}\mathrm{t})]\right\}\end{aligned} \tag{3.17}$$

where $\eta_{\mathrm{iR,L}} = \mu_0\omega_{\mathrm{iR,L}}/\mathrm{k_0}$ are the intrinsic impedances of the plasma for the respective modes.

A space charge current density, $\mathbf{J} = -\mathrm{eN_0}\mathbf{v_e}$, associated with the fields (3.16) and (3.17) will be induced in plasma. Their relationship is governed by the Ampere's law, $\mathbf{J} = -\epsilon_0\partial_\mathrm{t}\mathbf{E} + \nabla \times \mathbf{H}$. With the aid of (3.16) and (3.17), this induced current density is obtained to be

$$\begin{aligned}\mathbf{J}_{\pm}(\mathrm{z},\,\mathrm{t}) &= -\epsilon_0\sum_{\mathrm{i}=1}^{2}\left\{\left(\omega_{\mathrm{iR,L}} - \frac{\omega_0^2}{\omega_{\mathrm{iR,L}}}\right)E_{iR,L}^{+}\right. \\ &\quad \times[\hat{\mathbf{x}}\sin(\mathrm{k_0z}-\omega_{\mathrm{iR,L}}\mathrm{t}) \pm \hat{\mathbf{y}}\cos(\mathrm{k_0z}-\omega_{\mathrm{iR,L}}\mathrm{t})] \\ &\quad -\left(\omega_{\mathrm{iL,R}} - \frac{\omega_0^2}{\omega_{\mathrm{iL,R}}}\right)E_{iL,R}^{-} \\ &\quad \left.\times\,[\hat{\mathbf{x}}\sin(\mathrm{k_0z}+\omega_{\mathrm{iL,R}}\mathrm{t}) \pm \hat{\mathbf{y}}\cos(\mathrm{k_0z}+\omega_{\mathrm{iL,R}}\mathrm{t})]\right\}\end{aligned} \tag{3.18}$$

Again, for the same reason that it takes time for plasma to response to the wave fields as pointed out in Sec. 3.2, no current is generated at $t = 0^+$, i.e., $\mathbf{J}_\pm(z, t = 0^+) = 0$; thus, (3.18) leads to

$$\sum_{i=1}^{2}\left\{\left(\omega_{iR,L} - \frac{\omega_0^2}{\omega_{iR,L}}\right) E^+_{iR,L} - \left(\omega_{iL,R} - \frac{\omega_0^2}{\omega_{iL,R}}\right) E^-_{iL,R} = 0 \quad (3.19)$$

Applying the continuity conditions (3.15a) and (3.15b) on (3.16) and (3.17), yields

$$\sum_{i=1}^{2}(E^+_{iR,L} + E^-_{iL,R}) = E_0 \quad (3.20a)$$

and

$$\sum_{i=1}^{2}\left(\frac{1}{\omega_{iR,L}}E^+_{iR,L} - \frac{1}{\omega_{iL,R}}E^-_{iL,R}\right) = \frac{1}{\omega_0}E_0. \quad (3.20b)$$

With the aid of (3.20b), (3.19) reduces to

$$\sum_{i=1}^{2}(\omega_{iR,L}E^+_{iR,L} - \omega_{iL,R}E^-_{iL,R}) = \omega_0 E_0. \quad (3.20c)$$

Normalize the mode frequency by $\omega_0 = k_0 c$ and set $a_1 = \omega_{1R}/\omega_0, a_2 = \omega_{2R}/\omega_0$, and $a_3 = \omega_{1L}/\omega_0$, where a_1, a_2, and a_3 are determined by the dispersion equations (3.12) with $kc = \omega_0$. a_1 and a_2 are the two positive solutions (with $a_1 > a_2$) of the equation $a^3 - \beta a^2 - (1+\alpha)a + \beta = 0$; and a_3 is the positive solution of the equation $a^3 + \beta a^2 - (1+\alpha)a - \beta = 0$; here $\alpha = \omega_p^2/\omega_0^2$ and $\beta = \Omega_e/\omega_0$. Eqs. (3.20a) to (3.20c) are solved to obtain the electric field amplitude of each converted wave at $t = 0^+$.

1) In the RHCP case, the field amplitudes E^+_{1R} and E^+_{2R} of the two forward waves (R-mode and slow R-mode at frequencies ω_{1R} and ω_{2R}) and the field amplitude E^-_{1L} of the backward wave (L-mode at frequency ω_{1L}) are obtained to be

$$E^+_{1R} = -\left[\frac{a_1(a_3+1)(a_2-1)}{(a_3+a_1)(a_1)-a_2)}\right] E_0 \quad (3.21a)$$

$$E^+_{2R} = \left[\frac{a_2(a_3+1)(a_1-1)}{(a_3+a_2)(a_1-a_2)}\right] E_0 \quad (3.21b)$$

$$E^-_{1L} = -\left[\frac{a_3(a_1-1)(a_2-1)}{(a_3+a_2)(a_3+a_1)}\right] E_0. \quad (3.21c)$$

In this case, there are two transmitted waves and one reflected wave; the sum of the reflection coefficient Γ_{R} and the transmission coefficient $\mathrm{T_R}$ equals to 1, i.e., $\Gamma_{\mathrm{R}} + \mathrm{T_R} = 1$, where $\Gamma_{\mathrm{R}} = \Gamma_{\mathrm{R1L}} = E_{1L}^{-}/E_0$, and $\mathrm{T_R} = \mathrm{T_{R1R}} + \mathrm{T_{R2R}}$ with $\mathrm{T_{R1R}} = E_{1R}^{+}/E_0$ and $\mathrm{T_{R2R}} = E_{2R}^{+}/E_0$.

2) In the LHCP case, the field amplitude E_{1L}^{+} of the forward wave (L-mode at frequency $\omega_{1\mathrm{L}}$) and the field amplitudes E_{1R}^{-} and E_{2R}^{-} of the two backward waves (R-mode and slow R-mode at frequencies ω_{1R} and $\omega_{2\mathrm{R}}$) are obtained to be

$$E_{1L}^{+} = \left[\frac{\mathrm{a_3(a_1+1)(a_2+1)}}{\mathrm{(a_3+a_2)(a_3+a_1)}}\right] E_0 \tag{3.22a}$$

$$E_{1R}^{-} = \left[\frac{\mathrm{a_1(a_3-1)(a_2+1)}}{\mathrm{(a_3+a_1)(a_1-a_2)}}\right] E_0 \tag{3.22b}$$

$$E_{2R}^{-} = -\left[\frac{\mathrm{a_2(a_3-1)(a_1+1)}}{\mathrm{(a_3+a_2)(a_1-a_2)}}\right] E_0. \tag{3.22c}$$

In this case, there are one transmitted wave and two reflected waves; the sum of the reflection coefficient Γ_{L} and the transmission coefficient $\mathrm{T_L}$ equals to 1, i.e., $\Gamma_{\mathrm{L}} + \mathrm{T_L} = 1$, where $\Gamma_{\mathrm{L}} = \Gamma_{\mathrm{L1R}} + \Gamma_{\mathrm{L2R}}$, with $\Gamma_{\mathrm{L1R}} = E_{1\mathrm{R}}^{-}/E_0$ and $\Gamma_{\mathrm{L2R}} = E_{2\mathrm{R}}^{-}/E_0$ and $\mathrm{T_L} = E_{1\mathrm{L}}^{+}/E_0$.

The results presented in (3.21) and (3.22) should be checked with those of unmagnetized case. As $\beta = \Omega_{\mathrm{e}}/\mathrm{kc} \to 0$, $\mathrm{x_{2R}}$ in Fig. 3.2a also approaches to zero. Thus, in the unmagnetized situation, $\mathrm{a_2} = 0$ and $\mathrm{a_1} = \mathrm{a_3} = \omega/\omega_0$, where $\omega = \sqrt{\omega_{\mathrm{p}}^2 + \omega_0^2}$. The results in (3.21) and (3.22) reduce to

$$E_{1R}^{+} = E_{1L}^{+} = {}^1\!/\!{}_2\left(1 + \frac{\omega_0}{\omega}\right) E_0 = E_1^{+},$$

$$E_{1L}^{-} = E_{1R}^{-} = {}^1\!/\!{}_2\,(1 - \omega_0/\omega)E_0 = E_1^{-},$$

$$E_{2R}^{+} = E_{2R}^{-} = 0, \quad \text{and} \quad E_{2R}^{\pm}/\eta_{2R} = \pm[1 - (\omega_0/\omega)^2]E_0/\eta_0.$$

In the case of linear polarization,

$$\begin{aligned}\boldsymbol{E}(\mathrm{z}, \mathrm{t} \le 0) &= {}^1\!/\!{}_2\,[\boldsymbol{E}_+(\mathrm{z}, \mathrm{t} \le 0) + \boldsymbol{E}_-(\mathrm{z}, \mathrm{t} \le 0)] \\ &= \hat{\mathbf{x}}\, E_0 \cos(\mathrm{k_0 z} - \omega_0 \mathrm{t})\end{aligned}$$

and

$$\begin{aligned}\boldsymbol{H}(\mathrm{z}, \mathrm{t} \le 0) &= {}^1\!/\!{}_2\,[\boldsymbol{H}_+(\mathrm{z}, \mathrm{t} \le 0) + \boldsymbol{H}_-(\mathrm{z}, \mathrm{t} \le 0)] \\ &= \hat{\mathbf{y}}(E_0/\eta_0) \cos(\mathrm{k_0 z} - \omega_0 \mathrm{t});\end{aligned}$$

with the aid of (3.16) and (3.17), $\boldsymbol{E}$ (z, t > 0) and $\boldsymbol{H}$ (z, t > 0) are obtained to be

$$\begin{aligned}\boldsymbol{E}(z, t > 0) &= 1/2\,[\boldsymbol{E}_+(z, t > 0) + \boldsymbol{E}_-(z, t > 0)] \\ &= \hat{\mathbf{x}}\,1/2\,E_0\left[\left(1 + \frac{\omega_0}{\omega}\right)\cos(\mathrm{k_0 z} - \omega \mathrm{t})\right. \\ &\quad \left. + \left(1 - \frac{\omega_0}{\omega}\right)\cos(\mathrm{k_0 z} + \omega \mathrm{t})\right]\end{aligned} \tag{3.23a}$$

and

$$\begin{aligned}\boldsymbol{H}(\mathrm{z, t} > 0) &= 1/2\,[\boldsymbol{H}_+(\mathrm{z, t} > 0) + \boldsymbol{H}_-(\mathrm{z, t} > 0)] \\ &= \hat{\mathbf{y}}\,1/2\left(\frac{E_0}{\eta_0}\right)\left(\frac{\omega_0}{\omega}\right)\left[\left(1 + \frac{\omega_0}{\omega}\right)\cos(\mathrm{k_0 z} - \omega \mathrm{t})\right. \\ &\quad \left. - \left(1 - \frac{\omega_0}{\omega}\right)\cos(\mathrm{k_0 z} + \omega \mathrm{t})\right] + \hat{\mathbf{y}}\left(\frac{E_0}{\eta_0}\right)\left[1 - \left(\frac{\omega_0}{\omega}\right)^2\right]\cos \mathrm{k_0 z}.\end{aligned} \tag{3.23b}$$

These results are consistent with the previous results analyzed directly in the unmagnetized case. The second term on the right-hand side of (3.23b) is a wiggler magnetic field, which is converted from the whistler wave when the background magnetic field approaches zero.

3.3.3 *Momentum and energy conservation*

The power flux (Poynting vector) $\mathbf{S} = \mathbf{E} \times \mathbf{H}$, at $\mathrm{t} = 0^-$ and at $\mathrm{t} = 0^+$, are evaluated, with the aid of (3.14a) and (3.14b), at $\mathrm{t} = 0^-$,

$$\boldsymbol{E}_\pm(z, t = 0^-) = E_0(\hat{\mathbf{x}}\cos \mathrm{k_0 z} \mp \hat{\mathbf{y}}\sin \mathrm{k_0 z}),$$

and

$$\boldsymbol{H}_\pm(z, t = 0^-) = \pm(E_0/\eta_0)(\hat{\mathbf{x}}\sin \mathrm{k_0 z} \pm \hat{\mathbf{y}}\cos \mathrm{k_0 z}),$$

We then obtain

$$\mathbf{S}_< = \hat{\mathbf{z}}\frac{\mathrm{E}_0^2}{\eta_0}. \tag{3.24a}$$

With the aid of (3.16) and (3.17), at $\mathrm{t} = 0^+$,

$$\boldsymbol{E}_\pm(\mathrm{z, t} = 0^+) = \sum_{i=1}^{2}(E^+_{iR,L} + E^-_{iL,R})(\hat{\mathbf{x}}\cos \mathrm{k_0 z} \mp \hat{\mathbf{y}}\sin \mathrm{k_0 z})$$

and

$$\mathbf{H}_{\pm}(\mathrm{z},\,\mathrm{t}=0^{+}) = \pm\sum_{i=1}^{2}\{[(\omega_0/\omega_{\mathrm{iR,L}})E^{+}_{iR,L} - (\omega_0/\omega_{\mathrm{iL,R}})E^{-}_{iL,R}]/\eta_0\} \times(\hat{\mathbf{x}}\sin \mathrm{k}_0\mathrm{z} \pm \hat{\mathbf{y}}\cos \mathrm{k}_0 z)$$

we then obtain

$$\mathbf{S}_{>} = \hat{\mathbf{z}}\sum_{i=1}^{2}\sum_{j=1}^{2}(E^{+}_{iR,L} + E^{-}_{iL,R}) \times\left[\left(\frac{\omega_0}{\omega_{jR,L}}\right)E^{+}_{jR,L} - \left(\frac{\omega_0}{\omega_{jL,R}}\right)E^{-}_{jL,R}\right]/\eta_0. \tag{3.24b}$$

With the aid of (3.20a) and (3.20b), it shows that

$$\mathbf{S}_{<} = \mathbf{S}_{>},$$

attributed to the power conservation.

At $\mathrm{t} = 0^{-}$, the wave energy density is given to be

$$\begin{aligned} U(t=0^{-}) &= {}^{1}\!/\!{}_{2}\,\epsilon_0\left|\mathbf{E}_{\pm}(z,\,t=0^{-})\right|^{2} + {}^{1}\!/\!{}_{2}\,\mu_0\left|\mathbf{H}_{\pm}(z,\,t=0^{-})\right|^{2} \\ &= {}^{1}\!/\!{}_{2}\,\epsilon_0 E_0^2 + {}^{1}\!/\!{}_{2}\,\mu_0(E_0/\eta_0)^2 = \epsilon_0 E_0^2. \end{aligned} \tag{3.25a}$$

At $\mathrm{t} = 0^{+}$, there is no field energy stores in the just created plasma yet; the wave energy density is given to be

$$\begin{aligned} U(t=0^{+}) &= {}^{1}\!/\!{}_{2}\,\epsilon_0\left|\mathbf{E}_{\pm}(z,\,t=0^{+})\right|^{2} + {}^{1}\!/\!{}_{2}\,\mu_0\left|\mathbf{H}_{\pm}(z,\,t=0^{+})\right|^{2} \\ &= {}^{1}\!/\!{}_{2}\,\epsilon_0\left[\sum_{i=1}^{2}(E^{+}_{iR,L} + E^{-}_{iL,R})\right]^{2} \\ &\quad + {}^{1}\!/\!{}_{2}\,\mu_0\left\{\sum_{i=1}^{2}[(\omega_0/\omega_{iR,L})E^{+}_{iR,L} - (\omega_0/\omega_{iL,R})E^{-}_{iL,R}]\right\}^{2}/\eta_0^2 \end{aligned} \tag{3.25b}$$

Again, with the aid of (3.20a) and (3.20b), it shows that

$$\mathrm{U}(\mathrm{t} = 0^{-}) = \mathrm{U}(\mathrm{t} = 0^{+}),$$

attributed to the energy conservation.

The energy density U and the Poynting vector **S** of the wave are related to be

$$U = \frac{|\mathbf{S}|}{v_g}$$

where $v_g = \partial_k \omega$ is the group velocity of the wave. In terms of photon energy $\hbar\omega$, $U = N\hbar\omega$, where N is the number of photons per unit volume. The "Abraham" expression for the photon momentum in a dielectric medium is given by $p_A = \hbar\omega/nc$, where $n = n(\omega) = \sqrt{\epsilon_r}$, is the index of refraction, and c is the speed of light in free space; the effects of plasma dispersion on photon momentum are neglected because monochromatic wave is considered.

Thus, the momentum density **P** of each wave is given by

$$\mathbf{P} = \frac{\mathbf{S}}{v_g nc}.$$

At $t = 0^-$, $v_g = c$ and $\mathbf{S}_< = \hat{\mathbf{z}}(E_0^2/\eta_0)$, giving $\mathbf{P}_< = \hat{\mathbf{z}}(E_0^2/\eta_0 c^2)$; at $t = 0^+$, there are three waves, two forward and one backward waves in the case of RHCP and one forward and two backward wave in the case of LHCP.

1) In the case of RHCP, the Poynting vectors and the momenta of the two forward waves are

$$\boldsymbol{S}^+_{>iR} = \hat{\mathbf{z}}\frac{{E^+_{iR}}^2}{\eta_{iR}} \quad \text{and} \quad \mathbf{P}^+_{>iR} = \hat{\mathbf{z}}\frac{{E^+_{iR}}^2}{\eta_{iR} v_{giR} n_{iR} c},$$

respectively; where $i = 1, 2$; and those of the backward wave are

$$\boldsymbol{S}^-_{>1L} = -\hat{\mathbf{z}}\frac{{E^-_{1L}}^2}{\eta_{1L}} \quad \text{and} \quad \mathbf{P}^-_{>1L} = -\hat{\mathbf{z}}\frac{{E^-_{1L}}^2}{\eta_{1L} v_{g1L} n_{1L} c}.$$

Thus, the total momentum density of the converted waves at $t = 0^+$ is given to be

$$\mathbf{P}_>(RHCP) = \hat{\mathbf{z}}\left(\frac{{E^+_{1R}}^2}{\eta_{1R} v_{g1R} n_{1R} c} + \frac{{E^+_{2R}}^2}{\eta_{2R} v_{g2R} n_{2R} c} - \frac{{E^-_{1L}}^2}{\eta_{1L} v_{g1L} n_{1L} c}\right)$$
$$= \hat{\mathbf{z}}\frac{a_1\alpha_{1R}{E^+_{1R}}^2 + a_2\alpha_{2R}{E^+_{2R}}^2 - a_3\alpha_{1L}{E^-_{1L}}^2}{\eta_0 c^2} \tag{3.26a}$$

2) In the case of LHCP, the Poynting vector and the momentum of the forward wave are

$$\mathbf{S}_{>1L}^{+} = \hat{\mathbf{z}}\frac{\mathrm{E}_{1L}^{+\,2}}{\eta_{1L}} \quad \text{and} \quad \mathbf{P}_{>1L}^{+} = \hat{\mathbf{z}}\frac{\mathrm{E}_{1L}^{+\,2}}{\eta_{1L}\mathrm{v}_{g1L}\mathrm{n}_{1L}\mathrm{c}};$$

and those of the backward waves are

$$\mathbf{S}_{>iR}^{-} = -\hat{\mathbf{z}}\frac{\mathrm{E}_{iR}^{-\,2}}{\eta_{iR}} \quad \text{and} \quad \mathbf{P}_{>iR}^{-} = -\hat{\mathbf{z}}\frac{\mathrm{E}_{iR}^{-\,2}}{\eta_{iR}\mathrm{v}_{giR}\mathrm{n}_{iR}\mathrm{c}};$$

where i = 1, 2. Thus, the total momentum density of the converted waves at $t = 0^{+}$ is given to be

$$\begin{aligned}\mathbf{P}_{>}(LHCP) &= \hat{\mathbf{z}}\left(\frac{E_{1L}^{+\,2}}{\eta_{1L}\mathrm{v}_{g1L}\mathrm{n}_{1L}\mathrm{c}} - \frac{E_{1R}^{-\,2}}{\eta_{1R}\mathrm{v}_{g1R}\mathrm{n}_{1R}\mathrm{c}} - \frac{E_{2R}^{-\,2}}{\eta_{2R}\mathrm{v}_{g2R}\mathrm{n}_{2R}\mathrm{c}}\right)\\ &= \hat{\mathbf{z}}\frac{\mathrm{a}_3\alpha_{1L}E_{1L}^{+\,2} - \mathrm{a}_1\alpha_{1R}E_{1R}^{-\,2} - \mathrm{a}_2\alpha_{2R}E_{2R}^{-\,2}}{\eta_0\mathrm{c}^2}\end{aligned} \tag{3.26b}$$

The index of reflection n, the intrinsic impedance η, and the group velocity $\mathrm{v_g}$, at each mode frequency, are given, respectively, to be

$$n_{iR,L} = \frac{\omega_0}{\omega_{iR,L}}, \eta_{iR,L} = \left(\frac{\omega_{iR,L}}{\eta_0}\right)\eta_0 = \frac{\eta_0}{n_{iR,L}}, \tag{3.27}$$

and

$$\mathrm{v}_{giR,L} = \frac{\left(\frac{\omega_0}{\omega_{iR,L}}\right)\mathrm{c}}{\alpha_{iR,L}},$$

where

$$\alpha_{iR,L} = \left[1 \pm {}^1\!/_2\left(\frac{\Omega_e\omega_{iR,L}}{\omega_p^2}\right)\left(1 - \frac{\omega_0^2}{\omega_{iR,L}^2}\right)^2\right].$$

With the aid of (3.21a) to (3.21c) and (3.22a) to (3.22c), it is shown that

$$\begin{aligned}&\left(\mathrm{a}_1\alpha_{1R}E_{1R}^{+\,2} + \mathrm{a}_2\alpha_{2R}E_{2R}^{+\,2} - \mathrm{a}_3\alpha_{1L}E_{1L}^{-\,2}\right) = E_0^2\\ &= (\mathrm{a}_3\alpha_{1L}E_{1L}^{+\,2} - \mathrm{a}_1\alpha_{1R}E_{1R}^{-\,2} - \mathrm{a}_2\alpha_{2R}E_{2R}^{-\,2})\end{aligned}$$

Therefore,

$$\mathbf{P}_{>}(\mathrm{RHCP}) = \mathbf{P}_{<}(\mathrm{RHCP}) \quad \text{and} \quad \mathbf{P}_{>}(\mathrm{LHCP}) = \mathbf{P}_{<}(\mathrm{LHCP});$$

conservation of momentum in the mode transition is verified.

Problems

P3.1. Propagation of electromagnetic (EM) wave in unmagnetized uniform plasma is governed by the wave equation (3.1). Consider a case that a right-hand circularly polarized wave exists everywhere in free space prior to the plasma creation at t = 0, the wave fields are

$$\begin{cases} \mathbf{E}(z, t \le 0) = E_0[\hat{\mathbf{x}}\cos(k_0 z - \omega_0 t) - \hat{\mathbf{y}}\sin(k_0 z - \omega_0 t)] \\ \mathbf{H}(z, t \le 0) = \left(\frac{E_0}{\eta_0}\right)[\hat{\mathbf{x}}\sin(k_0 z - \omega_0 t) + \hat{\mathbf{y}}\cos(k_0 z - \omega_0 t)] \end{cases} \tag{P3.1}$$

where $k_0 = \omega_0/c$ and $\eta_0 = \sqrt{\mu_0/\epsilon_0}$ is the intrinsic impedance of the free space. The plasma is modelled by (3.3), which is suddenly created at t = 0.

Find $\boldsymbol{E}$ (z, t) and $\boldsymbol{H}$ (z, t) in t > 0; the results should contain a forward and a backward propagating wave at an upshifted frequency, and a static wiggler magnetic field.

P3.2. Spatially Periodic Plasma; consider an ideal case that a periodic slab plasma, as shown in Fig. P3.1, is created abruptly at t = 0 and there is no plasma decay afterward. The plasma frequency is modeled as

$$\begin{cases} \omega_p^2(z,t) = 0 & \text{in} \quad t < 0 \\ \omega_p^2(z,t) = \omega_{p0}^2 \sum\limits_{n=-\infty}^{\infty} P_{d/2}(z - nL - \ell - d/2) & \text{in} \quad t > 0 \end{cases} \tag{P3.2}$$

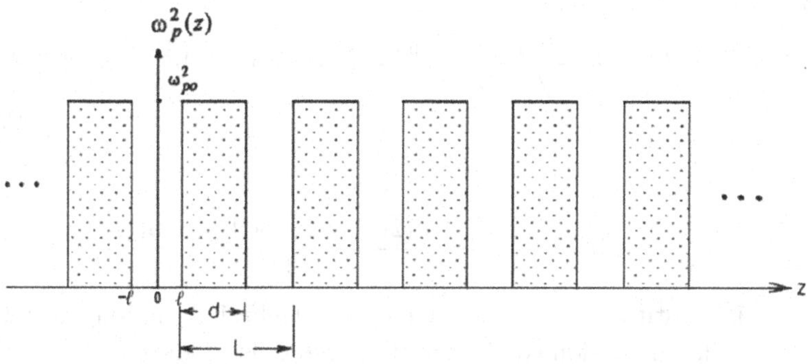

Figure P3.1. Geometry of the periodic plasma structure (after plasma creation) represented by the spatial distribution of the plasma density $\propto \omega_p^2(z)$.

where $P_{d/2}(z - a)$ is a unit rectangular pulse of width d centered about $z = a$, L is the separation between the centers of two adjacent slabs with $L \geq d$, $\ell = (L - d)/2$

For a time-harmonic wave with $\exp(-i\omega t)$ time dependence,

(1) Show that the phasor solution of Eq. (3.1) in t >0 for one spatial period $-\ell < z < \ell + d$, in Fig. P3.1, can be written as

$$E(z) = \begin{cases} A \exp(ikz) + B \exp(-ikz) & -\ell < z < \ell \\ C \exp[ik_1(z - \ell] + D \exp[-ik_1(z - \ell] & \ell < z < \ell + d \end{cases} \tag{P3.3}$$

where $k = \omega/c$ and $k_1 = \xi k$, and $\xi = \sqrt{1 - \omega_{p0}^2/\omega^2}$, are the wavenumbers in the free space and plasma slabs, and the index of refraction of the plasma slab.

(2) Apply the boundary conditions at $z = \pm\ell$: $E(\pm\ell^-) = E(\pm\ell^+)$ and $\partial_z E(\pm\ell^-) = \partial_z E(\pm\ell^-) = \partial_z E(\pm\ell^+)$ and using the Block wave condition:

$$E(z) = \exp(-i\beta L)E(z + L) \tag{P3.4}$$

to replace $E(-\ell^-)$ by $e^{-i\beta L}E[(\ell + d)^-]$, where $E[(\ell + d)^-]$ is given by (P3.3) and β is the propagation constant for the periodic structure as a whole, four algebraic equations are obtained. Show that B, C, D can be in terms of A as follows:

$$\begin{aligned} B &= -(b - \sqrt{b^2 - 1})A \\ C &= \left(\frac{1}{2\eta}\right)\left[(\xi + 1)e^{ik\ell} - (\xi - 1)(b - \sqrt{b^2 - 1})e^{-ik\ell}\right] A \\ D &= \left(\frac{1}{2\eta}\right)\left[(\xi - 1)e^{ik\ell} - (\xi + 1)(b - \sqrt{b^2 - 1})e^{-ik\ell}\right] A \end{aligned} \tag{P3.5}$$

where $\eta = \eta_0/\xi$ and

$$b = \frac{(1 + \xi^2)\cos(2k\ell) + 2\xi\cot(k_1 d)\sin(2k\ell)}{\xi^2 - 1}.$$

(3) Substitute (P3.5) into the fourth algebraic equation, show that the dispersion equation of the system is given to be

$$\cos\beta L = \cos(k_1 d)\cos 2k\ell - \left[\frac{\xi^2 + 1}{2\xi}\right]\sin(k_1 d)\sin 2k\ell \tag{P3.6}$$

(4) Let $E(z, t = 0^-) = E_0 \cos \mathrm{k_0 z}$, this field distribution does not change at $\mathrm{t} = 0^+$, i.e., $E(z, t = 0^+) = E_0 \cos \mathrm{k_0 z}$, because the plasma is created instantaneously. The phasor of this field is given by $E(z) = E_0 \exp(-\mathrm{i k_0 z})$. With the aid of the Block wave condition (P3.4), show that $\beta + k_0 = 2\pi \mathrm{p/L}$, where p is an integer.

(5) Since $\cos \beta \mathrm{L}$ is an even periodic function, (P3.6) has to be solved only for $0 \leq \beta \mathrm{L} \leq \pi$, i.e., $\mathrm{p} = 1$. Plot the dispersion curves in the case of $\mathrm{L} = 0.6\lambda_0$, $\mathrm{d} = 0.2\lambda_0$, and $\omega_{\mathrm{p0}} = 1.2\omega_0$, where λ_0 and $\mathrm{f_0} = \omega_0/2\pi$ are the wavelength and frequency of the original wave in free space. The frequency gap between two adjacent branches of the dispersion curves forms a stop band, or a band gap, which is one of the characteristic features of the periodic structures.

Chapter 4

Slow Varying Systems (One Dimensional Lumped Systems)

4.1 Introduction

In linear systems we have noted the characteristic role of single mode type equations

$$\left[\frac{\partial}{\partial t} + i\alpha\left(\frac{\nabla}{i}\right)^n\right]\Psi(\mathbf{r},t) = 0,$$

For n $= 1, 2$, and 3, which admit plane wave solutions of the form

$$\Psi = a e^{\mathrm{i}(\mathbf{k}\cdot\mathbf{r}-\omega \mathbf{t})} \equiv a e^{\mathrm{i}\theta(\mathbf{r},\mathbf{t})} \tag{4.1}$$

where in a homogeneous stationary medium the temporal frequency $\omega = -\partial\theta/\partial t$, the spatial vector wavenumber $\mathbf{k} = \nabla\theta$, and the amplitude a are all constant. It was pointed out that in a weakly inhomogeneous and even nonlinear medium it is possible to seek a single mode type solution of the same form, but where

$$a = a(\mathbf{r},t) \quad \text{and} \quad \mathbf{k} = \boldsymbol{\kappa}(\mathbf{r};\omega)$$

are functions of $\mathbf{r}, t$. If they vary slowly during one period of θ, these macroscopic quantities are of engineering interest. Furthermore, if medium is also nonstationary, then

$$a = a(\mathbf{r},t), \quad \mathbf{k} = \boldsymbol{\kappa}(\mathbf{r},t), \quad \text{and} \quad \omega = \nu(\mathbf{r},t)$$

will be set.

In reflection symmetric systems, we could equally well consider a basic two mode type wave equation

$$\left[\frac{\partial^2}{\partial t^2} + \beta\left(\frac{\nabla}{i}\right)^n\right]\Psi(\mathbf{r},\mathrm{t}) = 0,$$

which has solutions of the form

$$\frac{a}{2}e^{i\theta} \pm \frac{a}{2}e^{-i\theta} \quad \text{or} \quad \begin{cases} a\cos\theta \\ a\sin\theta \end{cases}$$

and variants of this equation for weakly inhomogeneous and nonstationary media.

Before discussing space-time applications of this type (4.1), and Whitham's average Lagrangian and Hamiltonian approach to finding the defining equations for $a, \boldsymbol{\kappa}, \nu$, it is worthwhile to consider similar types of problems in an one dimensional system, for example, lumped systems where variables are functions of time only. It will be similar to mode method discussed in Sec. 2.6 of Chapter 2, where the independent variable is "z".

In lumped systems, it is convenient to consider the oscillatory equation, which has a similar differential equation form as the Helmholtz equation (2.1),

$$\left(\frac{\mathrm{d}^2}{\mathrm{dt}^2} + \Omega^2\right)\mathrm{q(t)} = 0, \tag{4.2}$$

which, if Ω = constant, has solutions of the form

$$q(t) = \begin{cases} a\sin\theta \\ a\cos\theta \end{cases} \tag{4.3}$$

with

$$\theta = \omega t, \quad \text{a and } \omega \text{ are constants,} \quad \text{and} \quad \omega = \Omega.$$

Multiply $\dot{q} = dq/dt$ to (4.2), a "constant of the motion" is inferred by the resulting equation

$$\frac{d}{dt}\left(\frac{\dot{q}^2}{2} + \frac{\Omega^2}{2}q^2\right) = 0, \tag{4.4}$$

which implies that an equivalent unit mass particle of position $q(t)$ in a potential well

$$\mathrm{V}(q) = {}^1\!/\!{}_2\Omega^2 q^2,$$

its energy

$$H = {}^1\!/\!{}_2(\dot{q}^2 + \Omega^2 q^2) = {}^1\!/\!{}_2\Omega^2 a^2$$

is constant with time, where the two terms $1/2\dot{q}^2$ and $1/2\Omega^2 q^2$are the kinetic energy and potential energy of this particle, exchanging in this potential well. In terms of the initial position $q(0)$, the solution of (4.2) is given to be

$$q(t) = q(0)\cos\Omega t + \sqrt{a^2 - q^2(0)}\sin\Omega \mathrm{t} \tag{4.5}$$

4.2 Initial value problem for a one-dimensional lumped system-Duffing equation

An interesting class of nonlinear lumped system problems is defined by an undamped and unforced Duffing equation

$$\frac{\mathrm{d}^2 q}{\mathrm{dt}^2} + \Omega^2 q - 2\alpha^2 q^3 = 0, \tag{4.6}$$

where $q = q(t)$ is subject to the initial conditions, $q(0) = 0$, $\dot{q}(0) = \mathrm{v}_0$. Consider separately the cases:

(a) linear: $\Omega =$ constant, $\alpha = 0$; Eq. (4.6) reduces to (4.2)
(b) linear: $\Omega = \Omega(t)$, $\alpha = 0$
(c) nonlinear: $\Omega = \Omega(t)$, $\alpha \neq 0$

Case (a) Assume the ansatz:

$$q(t) = a\sin\theta,$$

where $\mathrm{d}\theta/\mathrm{dt} = \omega$. In other cases one can also use $q(t) = a\cos\theta$, or more generally, $q(t) = a_0 + a_1\sin\theta + a_2\cos\theta + \cdots$; the choice depends upon the initial conditions. On substitution into (4.6), defining equations for a and θ follow from

$$\frac{\mathrm{d}^2 q}{\mathrm{dt}^2} = \left(\frac{\mathrm{d}^2 a}{\mathrm{dt}^2} - a\omega^2\right)\sin\theta + \left(2\frac{\mathrm{d}a}{\mathrm{dt}}\omega + a\frac{d\omega}{dt}\right)\cos\theta,$$

as

$$\left(\frac{\mathrm{d}^2 a}{\mathrm{dt}^2} - a\omega^2 + a\Omega^2\right)\sin\theta + \left(2\frac{\mathrm{d}a}{\mathrm{dt}}\omega + a\frac{d\omega}{dt}\right)\cos\theta = 0, \tag{4.7}$$

whence, after setting the coefficients of the $\sin\theta$ and $\cos\theta$ terms equal to zero (since they are independent-the initial phase is arbitrary), we find

$$\omega^2 = \Omega^2 + \frac{1}{a}\frac{\mathrm{d}^2 a}{\mathrm{dt}^2} \quad \text{and} \quad \frac{d}{dt}(a^2\omega) = 0. \tag{4.8}$$

Thus, (4.6) is satisfied by $a(t) = a =$ constant and $\omega = \Omega =$ constant, where $\theta = \omega t$. Because $\dot{q}(t) = \Omega a\cos\Omega \mathrm{t}$, $\dot{q}(0) = \Omega a = \mathrm{v}_0$, the solution is obtained to be $q(t) = (\mathrm{v}_0/\Omega)\sin\Omega\mathrm{t}$, which is also given by (4.5) with $q(0) = 0$ and $a = \mathrm{v}_0/\Omega$.

Case (b) Assume the ansatz:

$$q(t) = a(t)\sin\theta,$$

where $\mathrm{d}\theta/\mathrm{dt} = \omega(t)$. One obtains the same defining equations (4.7) for $a(\mathrm{t})$ and $\omega(\mathrm{t})$; $a, \omega =$ constant is no longer a possible solution. If $\Omega(\mathrm{t})$ is weakly variable, then so also is ω, and (hence from (4.8)) so also is a. Thus, to a good approximation, one can set

$$\frac{1}{a}\frac{\mathrm{d}^2 a}{\mathrm{dt}^2} \cong 0,$$

and hence

$$\omega = \pm\Omega(t) \quad \textit{and} \quad a^2(t)\Omega(t) = a_0^2\Omega_0, \tag{4.9a}$$

$$\theta(t) = \int_0^t \Omega(t')dt', \tag{4.9b}$$

$$a^2(\mathrm{t}) = a_0^2\frac{\Omega_0}{\Omega(\mathrm{t})}, \tag{4.9c}$$

thus,

$$q(t) = (\mathrm{v}_0/\sqrt{\Omega_0\Omega})\sin\left(\int_0^t \Omega(\mathrm{t}')\,\mathrm{dt}'\right),$$

where $\Omega_0 = \Omega(0)$, $a_0 = a(0) = \mathrm{v}_0/\Omega_0$. A more accurate procedure requires an expansion in some smallness parameter.

Case (c) Assume the ansatz:

$$q(t) = a(t)\sin\theta,$$

where $\mathrm{d}\theta/\mathrm{dt} = \omega(\mathrm{t})$.

The defining equations for a and ω become

$$\left(\frac{\mathrm{d}^2 a}{\mathrm{dt}^2} - a\omega^2 + a\Omega^2\right)\sin\theta + \left(2\frac{\mathrm{d}a}{\mathrm{dt}}\omega + a\frac{d\omega}{dt}\right)\cos\theta$$
$$- 2\alpha^2 a^3 \sin^3\theta = 0, \tag{4.10}$$

Since

$$\sin^3\theta = \frac{3}{4}\sin\theta - \frac{1}{4}\sin 3\theta$$

then assuming that the third harmonic term $\alpha^2 a^3 \sin 3\theta$ can be neglected (because it is small and rapidly varying), and setting the independent coefficients of the $\sin\theta$ and $\cos\theta$ terms equal to zero, one finds

$$\omega^2 = \Omega^2 - \frac{3}{2}\alpha^2 a^2 + \frac{1}{a}\frac{d^2 a}{dt^2}, \tag{4.11a}$$

$$\frac{d}{dt}(a^2\omega) = 0. \tag{4.11b}$$

Again, for a weak nonlinearity (weak must be clarified in terms of a small parameter expansion), one sets $\frac{1}{a}\frac{d^2a}{dt^2} \cong 0$, whence

$$\omega = \pm\sqrt{\Omega^2 - \frac{3}{2}\alpha^2 a^2}, \tag{4.12a}$$

$$\theta = \int_0^t \omega(t')\, dt', \tag{4.12b}$$

$$a^2(t) = \frac{a_0^2\Omega_0\sqrt{1 - \frac{3}{2}\frac{\alpha^2 a_0^2}{\Omega_0^2}}}{\Omega(t)\sqrt{1 - \frac{3}{2}\frac{\alpha^2 a^2(t)}{\Omega^2(t)}}}. \tag{4.12c}$$

where $a(t)$ is in implicit form.

In the case of constant Ω, i.e., $\Omega = \Omega_0$, (4.12c) and (4.12a) leads to

$$a(t) = a_0 = v_0/\omega \quad \text{and} \quad \omega = (1/\sqrt{2})(\Omega_0^2 + \sqrt{\Omega_0^4 - 6\alpha^2 v_0^2}\,)^{1/2};$$

the approximate solution of (4.6) is obtained to be

$$q(t) = \frac{v_0}{\omega}\sin\omega t \tag{4.12d}$$

which is valid under the assumption that $(\alpha v_0/\Omega_0^2)^2 \ll 1$.

It is of interest to observe, upon multiplication of (4.6) by $\dot{q}$, that

$$\frac{d}{dt}\left(\frac{\dot{q}^2}{2} + \frac{\Omega^2 q^2}{2} - \frac{\alpha^2 q^4}{2}\right) = 0, \quad \text{for constant } \Omega. \tag{4.13}$$

Therefore, the value $1/2(\dot{q}^2 + \Omega^2 q^2 - \alpha^2 q^4) = H = 1/2 v_0^2$ of the bracketed expression is conserved and represents the energy of an equivalent unit mass particle moving in the potential field $V(q) = 1/2(\Omega^2 q^2 - \alpha^2 q^4)$. A similar relation had been noted previously for the linear case (Eq. (4.4)). This is because both systems are stationary.

In the stationary situation, i.e., Ω and α are constants, and when the object is trapped in the potential well, i.e., $H \leq \frac{\Omega^4}{8\alpha^2}$, Eq. (4.6) has analytical mode solutions.

(1) In the case of $H < \frac{\Omega^4}{8\alpha^2} = \frac{\Omega^2 q_M^2}{4}$, a periodic mode solution is given to be

$$q(t) = q_2 \,\mathrm{sn}\left(\frac{q_1}{\sqrt{2} q_M}\Omega \mathrm{t}, \beta\right), \tag{4.14}$$

where "sn" is a Jacobi elliptic function, $q_M = \frac{\Omega}{\sqrt{2}\alpha}$, $\beta = \frac{q_2}{q_1}$, and

$$q_1 = q_M \left(1 + \sqrt{1 - \frac{4H}{\Omega^2 q_M^2}}\right)^{\frac{1}{2}},$$

$$q_2 = q_M \left(1 - \sqrt{1 - \frac{4H}{\Omega^2 q_M^2}}\right)^{\frac{1}{2}}.$$

In the case of $(\alpha \mathrm{v}_0/\Omega_0^2)^2 \ll 1$, $\beta \cong \alpha \mathrm{v}_0/\Omega_0^2 \ll 1$; (4.14) reduces to (4.12d).

(2) When $H = \frac{\Omega^4}{8\alpha^2} = \frac{\Omega^2 q_M^2}{4}$, $q_2 = q_M = q_1$, and $\beta = 1$; hence,

$$q(t) = q_M \tanh \frac{\Omega}{\sqrt{2}} \mathrm{t}. \tag{4.15}$$

This is an aperiodic mode.

The above ideas, independently of the particular one-dimensional equation being considered, will be rephrased with Whitham's average Lagrangian and Hamiltonian method, to be introduced in Chapter 5.

4.3 Source excited oscillatory problem (forced Duffing oscillator)

A harmonic source term is added to the oscillatory equation, the resulting problem is a forced Duffing oscillator defined by the equation

$$\frac{\mathrm{d}^2 q}{\mathrm{d}t^2} + \Omega^2 q - 2\alpha^2 q^3 = -h \cos \nu t, \tag{4.16}$$

with prescribed initial conditions $q(0) = 0$ and $\dot{q}(0) = \mathrm{v}_0$. Again, consider separately the cases:

(a) linear: $\Omega = \text{constant}$, $\alpha = 0$
(b) linear: $\Omega = \Omega(t)$, $\alpha = 0$
(c) nonlinear: $\Omega = \text{constant}$, $\alpha \neq 0$

Case (a) Assume the ansatz:

$$q(t) = a\sin\theta + b\cos\nu t,$$

where $\mathrm{d}\theta/\mathrm{dt} = \Omega$, a and b are constants. Thus, $\theta = \theta_0 + \Omega t$, where θ_0 will be determined via the initial conditions. On substitution the ansatz $q(\mathrm{t})$ into (4.16), defining equations for a, b, and θ follow from

$$\frac{\mathrm{d}^2 q}{\mathrm{dt}^2} = -a\Omega^2\sin\theta - b\nu^2\cos\nu t,$$

as

$$-b(\nu^2 - \Omega^2)\cos\nu t = -h\cos\nu t, \tag{4.17}$$

which leads to $\mathrm{b} = \mathrm{h}/(\nu^2 - \Omega^2)$. The initial conditions define $a\sin\theta_0 + b = 0$ and $a\Omega\cos\theta_0 = \mathrm{v}_0$, which lead to $\theta_0 = -\tan^{-1}(b\Omega/\mathrm{v}_0)$ and $a = -b/\sin\theta_0$.

When the source drives the oscillator at resonance, i.e., $\nu = \Omega$, the ansatz $q(\mathrm{t}) = a\sin\theta$ is assumed, where again, $\mathrm{d}\theta/dt = \Omega$, i.e., $\theta = \theta_0 + \Omega t$, but $a = a(t)$ is not a constant. On substitution into (4.16) and set $\nu t = \theta - \theta_0$ in the right-hand side term of (4.16), defining equations for a follow from

$$\frac{\mathrm{d}^2 q}{\mathrm{dt}^2} = \left(\frac{\mathrm{d}^2 a}{\mathrm{dt}^2} - a\Omega^2\right)\sin\theta + 2\frac{\mathrm{d}a}{\mathrm{dt}}\Omega\cos\theta,$$

as

$$\frac{\mathrm{d}^2 a}{\mathrm{dt}^2}\sin\theta + 2\frac{\mathrm{d}a}{\mathrm{dt}}\Omega\cos\theta = -h(\cos\theta_0\cos\theta + \sin\theta_0\sin\theta), \tag{4.18}$$

whence, after setting the coefficients of the $\sin\theta$ and $\cos\theta$ terms equal to zero, we find $\theta_0 = 0$ and $a(t) = (\mathrm{v}_0 - ht/2)/\Omega$.

Plot in Fig. 4.1 are the oscillatory responses in non-resonance case (A) with the parameters: $\Omega = 1$, $\nu = 2$, and $h = 0.2$ and in the resonance case (B) with the parameters: $\Omega = 1$, $\nu = 1$, and $h = 0.2$. In both cases, the initial conditions are $q(0) = 0$ and $\dot{q}(0) = 1$. As shown, after the transient, the resonance force drives the oscillation amplitude to increase linearly in time.

Case (b) Assume the ansatz:

$$q(\mathrm{t}) = a\sin\theta + b\cos(\phi + \nu t),$$

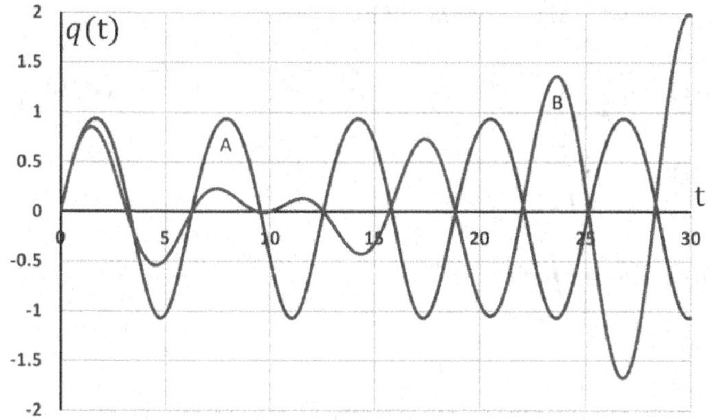

Figure 4.1. Linear oscillator with non-resonance and resonance sources; plots A and B are the oscillatory responses to a non-resonance and a resonance source, respectively.

where $\mathrm{d}\theta/\mathrm{dt} = \omega$, $\phi_0 = 0$, $\mathrm{d}\phi/\mathrm{dt} = \sigma(t)$, $a = a(t)$, and $b = b(t)$. On substitution into (4.16), defining equations for a, b, θ, and ϕ follow from

$$\frac{\mathrm{d}^2 q}{\mathrm{dt}^2} = \left(\frac{\mathrm{d}^2 a}{\mathrm{dt}^2} - \omega^2 a\right)\sin\theta + \left(2\frac{\mathrm{da}}{\mathrm{dt}}\omega + a\frac{d\omega}{dt}\right)\cos\theta$$
$$+ \left(\frac{\mathrm{d}^2 b}{\mathrm{dt}^2} - \nu_1^2 b\right)\cos(\phi + \nu t) - \left(2\frac{\mathrm{db}}{\mathrm{dt}}\nu_1 + b\frac{d\nu_1}{dt}\right)\sin(\phi + \nu t),$$

as

$$\left(\frac{\mathrm{d}^2 a}{\mathrm{dt}^2} - a\omega^2 + a\Omega^2\right)\sin\theta + \left(2\frac{\mathrm{da}}{\mathrm{dt}}\omega + a\frac{d\omega}{dt}\right)\cos\theta$$
$$+ \left(\frac{\mathrm{d}^2 b}{\mathrm{dt}^2} - \nu_1^2 b + \Omega^2 b\right)\cos(\phi + \nu t) - \left(2\frac{\mathrm{db}}{\mathrm{dt}}\nu_1 + b\frac{d\nu_1}{dt}\right)\sin(\phi + \nu t)$$
$$= -h\cos\nu t, \tag{4.19}$$

where $\nu_1 = \nu + \sigma(t)$.

Thus, to a good approximation in the early time, one sets

$$\frac{1}{a}\frac{\mathrm{d}^2 a}{\mathrm{dt}^2} \cong 0, \quad \frac{1}{b}\frac{\mathrm{d}^2 b}{\mathrm{dt}^2} \cong 0, \quad \text{and} \quad \phi \cong 0 \text{ (i.e., } \nu_1 \sim \nu),$$

and hence

$$\omega = \pm\Omega(\mathrm{t}), \quad \mathrm{a}^2(\mathrm{t})\Omega(\mathrm{t}) = \mathrm{a}_0^2\Omega_0, \quad \mathrm{b} = \mathrm{h}/(\nu_1^2 - \Omega^2),$$
$$\text{and} \quad \mathrm{b}^2(\mathrm{t})\nu_1(\mathrm{t}) = \mathrm{b}_0^2\nu, \tag{4.19a}$$

where $b_0 = h/(\nu^2 - \Omega_0^2)$ and $\Omega_0 = \Omega(0)$; hence,

$$\nu_1(\mathrm{t}) = \left(\frac{\nu}{2}\right)\left\{[1-(\Omega_0/\nu)^2] + \sqrt{[1-(\Omega_0/\nu)^2]^2 + 4(\Omega/\nu)^2}\right\},$$

$$b(\mathrm{t}) = b_0\sqrt{\frac{\nu}{\nu_1}},$$

$$\theta(t) = \theta_0 + \int_0^{\mathrm{t}} \Omega(\mathrm{t}')\mathrm{dt}', \tag{4.19b}$$

and

$$a^2(\mathrm{t}) = a_0^2\frac{\Omega_0}{\Omega(\mathrm{t})}, \tag{4.19c}$$

where $\theta_0 = -\tan^{-1}(b_0\Omega_0/\mathrm{v}_0)$ and $a_0 = -\mathrm{b}_0/\sin\theta_0$ are determined by the initial conditions as being demonstrated in case (a).

Consider an example with the parameters: $\Omega = 1 + 0.01\mathrm{t}$, $\nu = 2$, and $\mathrm{h} = 0.2$, and the same initial conditions $q(0) = 0$ and $\dot{q}(0) = 1$, and Eqs. (4.19) are applied to plot the oscillatory response of this oscillator. A direct integration of Eq. (4.16) is also performed. Both plots are presented in Fig. 4.2 for a close comparison. As shown, plot B, obtained from Eq. (4.19), matches very well with plot A, obtained from direct integration of Eq. (4.16).

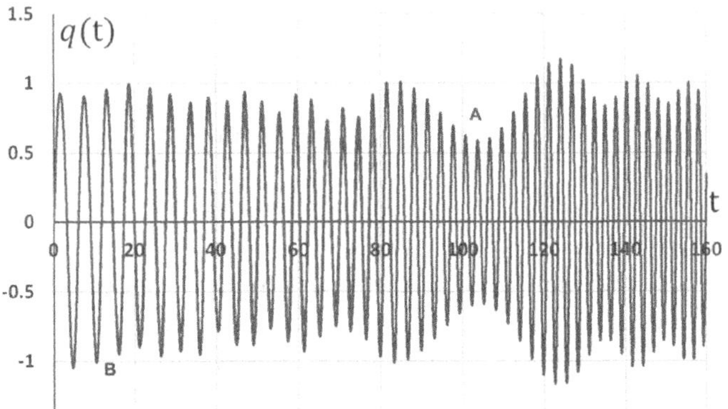

Figure 4.2. Linear oscillator with the resonance frequency increasing linearly in time and driven by a non-resonance source. Plot A is obtained from a direct integration of Eq. (4.16) and plot B is based on Eqs. (4.19).

Case (c) Consider the ansatz:

$$q(\mathrm{t}) = a(\mathrm{t})\cos[\nu t + \theta(t)],$$

which is useful for the regime $\nu \cong \Omega$. Substitute into (4.16), and equating coefficients of $\sin[\nu t + \theta(t)]$ and $\cos[\nu \mathrm{t} + \theta(t)]$ on both sides, results in

$$\frac{d}{dt}[a^2(\nu + \dot{\theta})] = ha\sin\theta, \tag{4.20a}$$

$$(\nu + \dot{\theta})^2 = \Omega^2 - \frac{3}{2}\alpha^2 a^2 + \frac{h}{a}\cos\theta + \frac{1}{a}\frac{d^2 a}{dt^2}. \tag{4.20b}$$

Expand Eqs. (4.20) to first order in $\dot{\theta}$ and $\dot{a}$ gives the approximate defining equations (with neglect of small, rapidly varying terms, i.e., $|\ddot{a}| \ll |h\cos\theta|$) for a and θ; first, (4.20b) becomes

$$\frac{\mathrm{d}\theta}{\mathrm{dt}} = \frac{\Omega^2 - \nu^2}{2\nu} - \frac{3\alpha^2 a^2}{4\nu} + \frac{h}{2a\nu}\cos\theta. \tag{4.21a}$$

Take a time derivative on both sides of (4.21a), yields

$$\frac{\mathrm{d}^2\theta}{\mathrm{dt}^2} = -\left(\frac{3\alpha^2 a}{2\nu} + \frac{h}{2a^2\nu}\cos\theta\right)\frac{\mathrm{da}}{\mathrm{dt}} - \frac{h}{2a\nu}\sin\theta\frac{\mathrm{d}\theta}{\mathrm{dt}}. \tag{4.21b}$$

With the aid of (4.21b), (4.20a) becomes

$$\frac{da}{dt} = \frac{h}{2\nu}\sin\theta\frac{\left(1 + \frac{d\theta/dt}{2\nu}\right)}{\left[1 - \frac{9}{8}\left(\frac{\alpha a}{\nu}\right)^2 + \frac{d\theta/dt}{2\nu}\right]}, \tag{4.21c}$$

Although Eqs. (4.21a) and (4.21c) seem no easier to solve than Eq. (4.16), the quantities of interest, a and θ, are slowly varying. Consequently, a numerical calculation can easily be carried out. This is in contrast to the situation for (4.16), where the possible high frequency variability of $q(\mathrm{t})$ often makes the numerical calculation intractable.

Consider a resonance driving example; this nonlinear oscillator has the parameters: $\Omega = 1$, $\nu = 1$, $\alpha^2 = 0.1$, and $\mathrm{h} = 0.2$, and the initial conditions $q(0) = 0$ and $\dot{q}(0) = 1$. The approximate defining equations (4.21a) and (4.21c) are solved for the oscillatory response of this nonlinear oscillator.

It is then closely compared with that from a direct integration of Eq. (4.16) by overlapping two plots. As shown in Fig. 4.3, plot B, obtained from Eqs. (4.21a) and (4.21c), matches quite well with plot A, obtained from direct integration of Eq. (4.16).

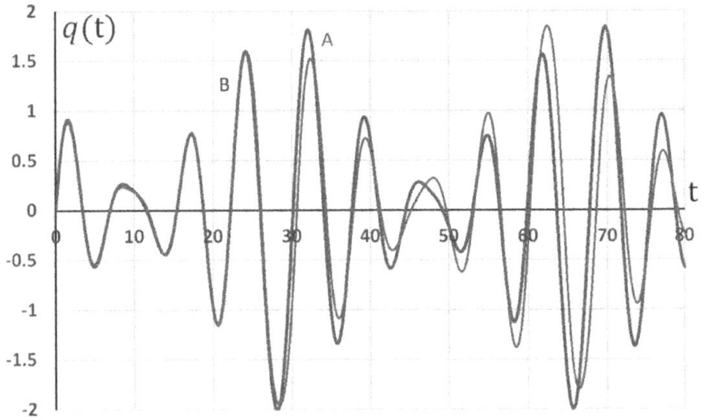

Figure 4.3. Nonlinear oscillator driven by a resonance source; plot B is based on Eqs. (4.21a) and (4.21c) and plot A is obtained from a direct integration of Eq. (4.16).

The mode approach illustrated above appears to be applicable for analyzing quasi-periodic oscillation in linear slow time-varying systems as well as in weak nonlinear systems.

4.4 Oscillatory problem with friction (one dimensional lumped systems with damping)

Next, a damped oscillator is considered. This one-dimensional system defined by

$$\frac{d^2q}{dt^2} + D\frac{dq}{dt} + Eq + Fq^2 + Gq^3 = 0, \tag{4.22}$$

subject to the prescribed initial conditions q(0) and $\dot{q}(0)$, provides interesting insights into different types of linear and nonlinear responses. Upon multiplication of (4.22) by $\dot{q}$, it becomes

$$\frac{d}{dt}\left(\frac{\dot{q}^2}{2} + \frac{E}{2}q^2 + \frac{F}{3}q^3 + \frac{G}{4}q^4\right) = -D\dot{q}^2 \tag{4.23}$$

This equation is descriptive of a unit object with position q(t) and velocity $\dot{q}$(t) = dq/dt, oscillating in a potential field specified by

$$U(q) = \frac{E}{2}q^2 + \frac{F}{3}q^3 + \frac{G}{4}q^4, \tag{4.23a}$$

with friction defined by a collision frequency D. Eq. (4.23) can be expressed to be

$$\frac{d}{dt}[T(\dot{q}) + U(q)] = F_c(\dot{q})\dot{q} \tag{4.23b}$$

where the kinetic energy $T(\dot{q}) = \dot{q}^2/2$ and the friction force $F_c(\dot{q}) = -D\dot{q}$. It is noted that (4.23b) is a power conservation equation $\frac{d}{dt}\varepsilon = -D\dot{q}^2 < 0$ for $D > 0$, where $\varepsilon = \mathrm{T} + \mathrm{U}$ is the total energy.

Computer displays of the response of this system for various parameters and initial conditions are depicted in Figs. 4.4–4.6. In all cases the (a) and (c) plots correspond to similar initial conditions: $q(0) = q_0$ and $\dot{q}(0) = 0$.

1. The simple linear case with the parameters $D = 0.05$, $E = 1$, $F = 0$, $G = 0$; it corresponds to a damped, constant frequency, oscillatory response, which is shown in Fig. 4.4. In this case, (4.22) can be solved analytically by the following procedure.

Assume the ansatz:

$$q(t) = ae^{-\alpha t}\cos\theta,$$

where $\mathrm{d}\theta/\mathrm{d}t = \Omega$, a, α, and Ω are constants. On substitution into (4.22), defining equations for α and Ω follow from

$$\frac{dq}{\mathrm{dt}} = -\Omega a\mathrm{e}^{-\alpha t}\sin\theta - \alpha a\mathrm{e}^{-\alpha t}\cos\theta,$$

and

$$\frac{\mathrm{d}^2 q}{\mathrm{dt}^2} = -(\Omega^2 - \alpha^2)a\mathrm{e}^{-\alpha t}\cos\theta + 2\alpha\Omega a\mathrm{e}^{-\alpha t}\sin\theta,$$

as

$$[-(\Omega^2 - \alpha^2) + 1]\cos\theta + (2\alpha\Omega - 0.05\Omega)\sin\theta = 0 \tag{4.24}$$

which leads to $\alpha = 0.025$ and $\Omega = \sqrt{1+\alpha^2}$. The initial conditions define $a\cos\theta_0 = q_0$ and $\Omega\sin\theta_0 + \alpha\cos\theta_0 = 0$, which lead to

$$\theta_0 = -\tan^{-1}\left(\frac{\alpha}{\Omega}\right) \quad \text{and} \quad a = \mathrm{q}_0/\cos\theta_0.$$

Hence,

$$q(t) = \mathrm{q}_0\sqrt{1 + (\alpha/\Omega)^2}\mathrm{e}^{-\alpha t}\cos(\theta_0 + \Omega t) \tag{4.24a}$$

(a) Figure 4.4(a) is a plot of $q(t)$ in (4.24a) versus t for various values of q_0.
(b) Figure 4.4(b) displays the potential $U(q) = \frac{1}{2}q^2$, given in (4.23a), versus q.

(c) Figure 4.4(c) is a phase plane plot of $\dot{q}(t)$ versus $q(t)$ for various $\dot{q}(0)$ with $q(0) = 0$.

The potential well shown in Fig. 4.4(b) has a minimum at $q = 0$; thus, the damped oscillation has an equilibrium at the point (0, 0) of the phase plane plot in Fig. 4.4(c).

2. A nonlinear case with $D = 0.05$, $E = 1$, $F = -1$, $G = 0$; it corresponds to a damped, oscillatory response with variable and amplitude dependent frequency, which is pictured in Fig. 4.5. It should be noted that the response is unstable for initial condition $q(0) > 1$.

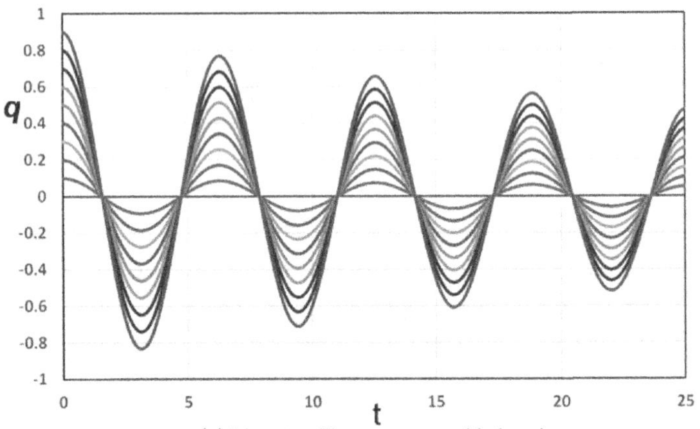

(a) Linear oscillator response with damping.

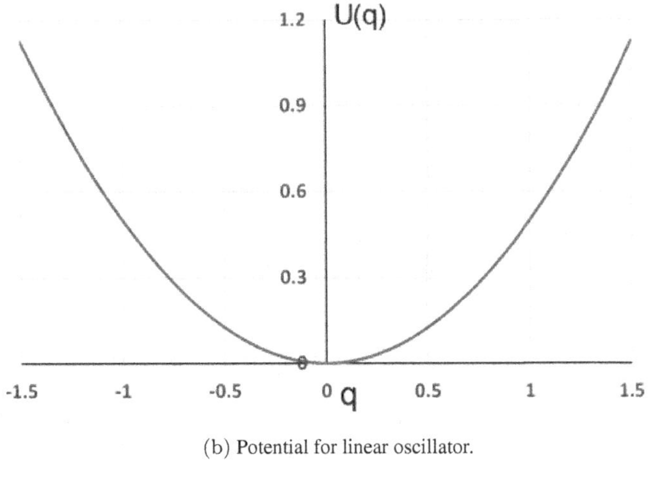

(b) Potential for linear oscillator.

Figure 4.4. A damped, constant frequency, oscillatory response of a linear oscillator.

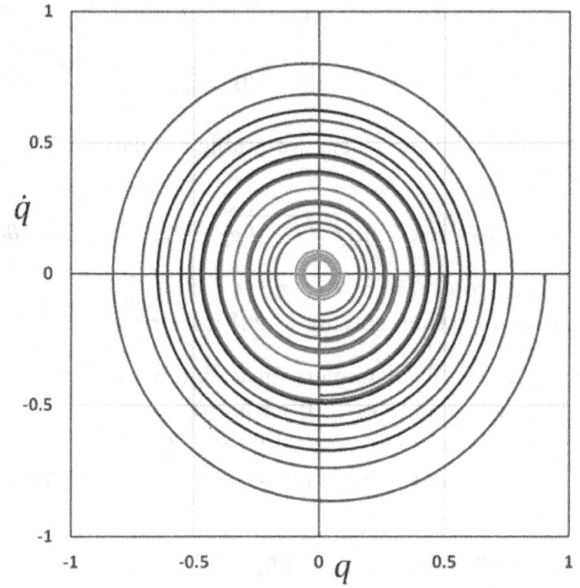

(c) Phase plane plot for linear oscillator with damping.

Figure 4.4. (*Continued*)

Assume the ansatz:

$$q(t) = \mathrm{b(t)} + a\mathrm{e}^{-\alpha t} \cos\theta,$$

where $\mathrm{d}\theta/\mathrm{dt} = \Omega$, a, α, and Ω are constants. On substitution into (4.22), defining equations for α and Ωfollow from

$$\frac{dq}{\mathrm{dt}} = -\Omega a\mathrm{e}^{-\alpha t} \sin\theta - \alpha a\mathrm{e}^{-\alpha t} \cos\theta,$$

and

$$\frac{\mathrm{d}^2 q}{\mathrm{dt}^2} = -(\Omega^2 - \alpha^2) a\mathrm{e}^{-\alpha t} \cos\theta + 2\alpha\Omega a\mathrm{e}^{-\alpha t} \sin\theta,$$

as

$$\begin{aligned}
&[-(\Omega^2 - \alpha^2) + 1 - 2b - 0.05\alpha] a\mathrm{e}^{-\alpha t} \cos\theta \\
&\quad + (2\alpha\Omega - 0.05\Omega) a\mathrm{e}^{-\alpha t} \sin\theta \\
&\quad + \left(\frac{\mathrm{d}^2\mathrm{b}}{\mathrm{dt}^2} + 0.05\frac{\mathrm{db}}{\mathrm{dt}} + \mathrm{b} - \mathrm{b}^2 - \frac{a^2}{2}\mathrm{e}^{-2\alpha t}\right) = 0; \qquad (4.25)
\end{aligned}$$

it leads to $\alpha = 0.025$ and $\Omega = \sqrt{1 + \alpha^2 - 2b - 0.05\alpha}$.

To a good approximation, one sets $\ddot{b} \cong 0$ and $0.05\dot{\mathrm{b}} \cong 0$, yields

$$b = \frac{(1 - \sqrt{1 - 2a^2 \mathrm{e}^{-2\alpha t}})}{2} \quad \text{and} \quad b(0) = b_0 = \frac{(1 - \sqrt{1 - 2a^2})}{2}$$

for a stable response; thus $a^2 \leq 1/2$ for stable oscillation. The initial conditions define

$$b_0 + a\cos\theta_0 = q_0 \quad \text{and} \quad a\alpha/\sqrt{1 - 2a^2} + \Omega\sin\theta_0 + \alpha\cos\theta_0 = 0,$$

which can be solved to obtain a, θ_0, and $\theta = \theta_0 + \int_0^t \Omega(t')dt'$. However, it seems no easier to solve than solving (4.22) numerically.

(a) Equation (4.22) is solved numerically. Figure 4.5(a) is a plot of q(t) versus t for various values of q(0) with $\dot{\mathrm{q}}(0) = 0$.
(b) Figure 4.5(b) displays the potential $U(q) = \frac{1}{2}q^2 - \frac{1}{3}q^3$ versus q.
(c) Figure 4.5(c) is a phase plane plot of $\dot{\mathrm{q}}(\mathrm{t})$ versus q(t) for various $\dot{\mathrm{q}}(0)$ with q(0) = 0.

The potential well shown in Fig. 4.5(b) has a maximum at q = 1 and a minimum at q = 0; thus, the response is also unstable if the initial energy $\varepsilon(0)$ is too large. For stable response, the damped oscillation has an equilibrium at the point (0, 0) of the phase plane plot in Fig. 4.5(c).

3. The nonlinear case with $D = 0.05$, $E = 0.5$, $F = -1.5$, $G = 1$; it corresponds to a damped, oscillatory, amplitude dependent and variable frequency response with two distinct stationary states.

On substitution the ansatz:

$$q(t) = \mathrm{b(t)} + a\mathrm{e}^{-\alpha t}\cos\theta$$

into (4.22), defining equations for α and Ω follow from

$$\begin{aligned}&[-(\Omega^2 - \alpha^2) + 0.5 - 3\mathrm{b} - 0.05\alpha + 3\mathrm{b}^2 + 0.75a^2\mathrm{e}^{-2\alpha t}]\\&\quad \times a\mathrm{e}^{-\alpha t}\cos\theta + (2\alpha\Omega - 0.05\Omega)a\mathrm{e}^{-\alpha t}\sin\theta\\&\quad + \left(\frac{\mathrm{d}^2\mathrm{b}}{\mathrm{dt}^2} + 0.05\frac{\mathrm{db}}{\mathrm{dt}} + 0.5\mathrm{b} - 1.5\mathrm{b}^2 - 0.75a^2\mathrm{e}^{-2\alpha t} + b^3 + 1.5ba^2\mathrm{e}^{-2\alpha t}\right) = 0,\end{aligned} \tag{4.26}$$

which leads to

$$\Omega = \sqrt{0.5 + \alpha^2 - 3b - 0.05\alpha + 3\mathrm{b}^2 + 0.75a^2\mathrm{e}^{-2\alpha t}}$$

and $\alpha = 0.025$.

Again, to a good approximation, one sets $\ddot{\text{b}} \cong 0$ and $0.05\dot{\text{b}} \cong 0$ to have

$$b^3 - 1.5\text{b}^2 + 0.5\text{b}(1 + 3a^2\text{e}^{-2\alpha\text{t}}) - 0.75a^2\text{e}^{-2\alpha\text{t}} = 0;$$

it leads to 3 stable oscillations, and two of them are in distinct stationary states (bistable states). The initial conditions define

$$b_0 + a\cos\theta_0 = q_0 \quad \text{and} \quad -\dot{\text{b}}_0/a + \Omega\sin\theta_0 + \alpha\cos\theta_0 = 0,$$

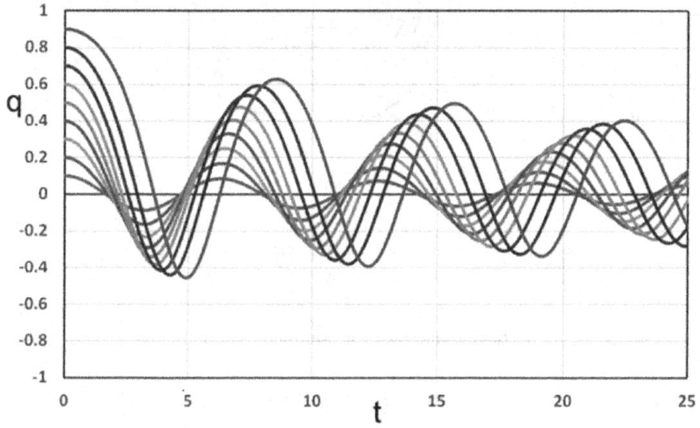

(a) Nonlinear oscillator response.

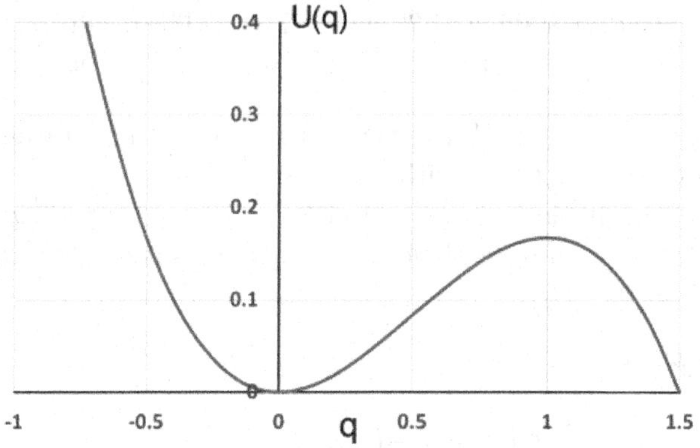

(b) Potential for nonlinear oscillator.

Figure 4.5. A damped, oscillatory response, with variable and amplitude dependent frequency, of a nonlinear oscillator.

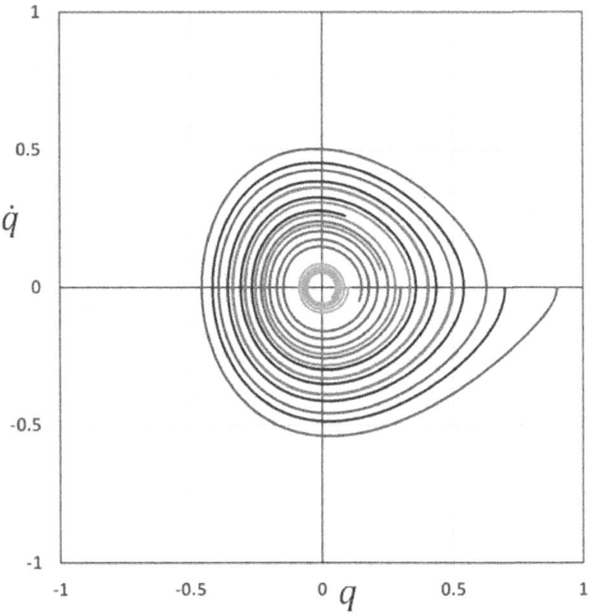

(c) Phase plane plot for nonlinear oscillator.

Figure 4.5. (*Continued*)

which can be solved to obtain a, θ_0, and $\theta = \theta_0 + \int_0^t \Omega(t')dt'$. Again, it seems no easier to solve than solving (4.22) numerically. The results obtained from the direct numerical integration of (4.22) are plotted in Fig. 4.6.

(a) Figure 4.6(a) is a plot of q(t) versus t in two distinct stationary states for various values of q(0) with $\dot{q}(0) = 0$.
(b) Figure 4.6(b) displays the potential $U(q) = \frac{1}{4}q^2 - \frac{1}{2}q^3 + \frac{1}{4}q^4$ versus q.
(c) Figure 4.6(c) is a phase plane plot of $\dot{q}(t)$ versus q(t) for various $\dot{q}(0)$ with $q(0) = 0$.

On setting,

$$\frac{d}{dq}U(q) = 0 = 0.5q(1 - 3q + 2q^2),$$

it leads to two minima at q = 0 and 1, and a maximum at q = 0.5 as shown in Fig. 4.6(b); thus, there are three equilibria in the phase plane plot of

Fig. 4.6(c); two stable ones are at the bottom of U(q) at (0,0) and (1, 0) and an unstable one at the peak of U(q) at (0.5, 0). In Fig. 4.6(c), all the initial conditions converge to one of the two stable equilibria. In the strong nonlinear situation considered in cases 2 and 3, the mode approach appears to be no easier to solve than the numerical approach.

In the next Chapter, a Lagrangian method is introduced. It invokes average procedures to remove the possible high frequency variability of the oscillatory response, which often makes the numerical calculation intractable.

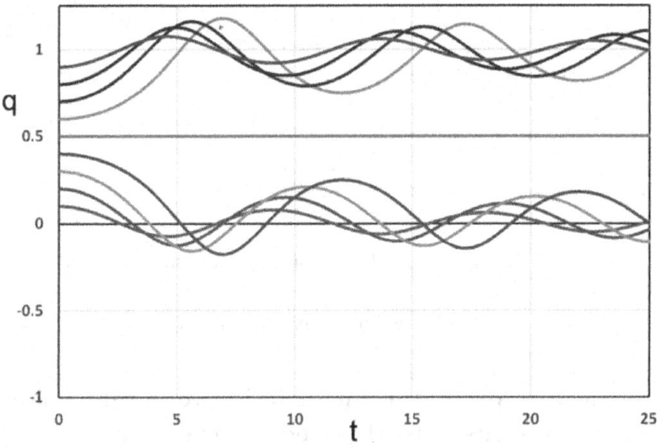

(a) Nonlinear oscillator response (bistable).

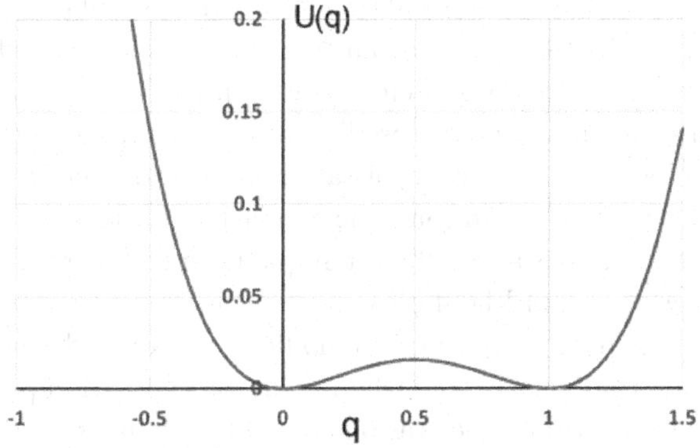

(b) Potential for nonlinear oscillator.

Figure 4.6. A damped, oscillatory, amplitude dependent and variable frequency response of a nonlinear oscillator with two distinct stationary states.

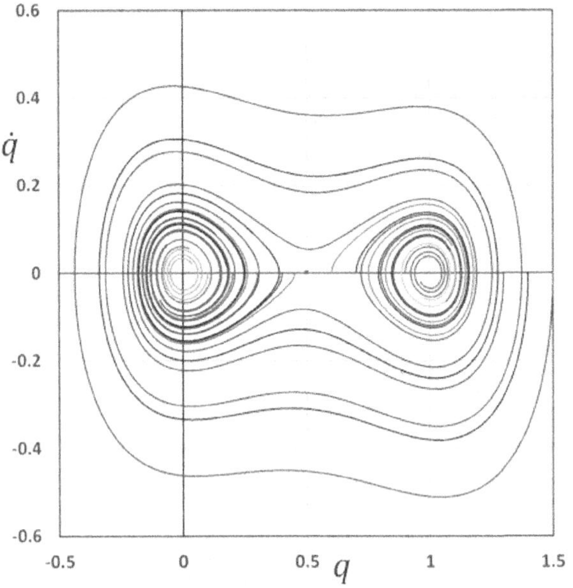

(c) Phase plane plot for nonlinear oscillator.

Figure 4.6. (*Continued*)

4.5 Forced bistate oscillator with friction (one dimensional nonlinear systems with three equilibria — from deterministic to chaotic)

In case (c) of Sec. 4.3, a forced single state nonlinear oscillator was studied. This oscillator is driven by a resonance force, which is counterbalanced by the nonlinear force of the system. As shown in Fig. 4.3, the oscillation amplitude varies in time, but the oscillation remains to be periodic.

In case 3 of Sec. 4.4, a damped bistate nonlinear oscillator was studied. Depending on the initial conditions, the oscillation is attracted to one of the two stable states and spirals in the phase plane to the corresponding equilibrium point as shown in Fig. 4.6(c). In this case, as shown in Fig. 4.6(a), the oscillation remains to be quasi-periodic. However, when a resonance force, similar to that on the right-hand side of (4.16), is applied, it will forbid the oscillator to become stationary. This system is descriptive with the equation

$$\frac{d^2q}{dt^2} + D\frac{dq}{dt} + Eq + Fq^2 + Gq^3 = -h\cos\nu t \tag{4.27}$$

One type of the potential function $U(q) = \frac{E}{2}q^2 + \frac{F}{3}q^3 + \frac{G}{4}q^4$, illustrated in Fig. 4.6b, contains two bounded wells and one unbounded one.

When the driven force can overcome the friction, the object oscillates in the potential wells and bifurcation of the trajectory may occur to show interesting insights of the bistate nonlinear responses. A further increase of the driven force, the oscillation may become aperiodic without regularity.

A progression toward dynamic chaos is presented via the following example in three cases. The parameters used in the example are $D = 0.05$, $E = (0.2\pi)^2$, $F = -1.5 \times (0.2\pi)^2$, $G = 0.5 \times (0.2\pi)^2$, $\nu = 0.2\pi$ for the three cases: (1) h = 0; (2) h = 0.05; and (3) h = 0.2. The potential function is presented in Fig. 4.7.

1. h = 0; Eq. (4.27) describes a damped nonlinear oscillator, which was presented in case 3 of Sec. 4.4. Any initial oscillation will be damped to settle at one of the two equilibria, depending on the initial conditions. This is demonstrated in Fig. 4.8.
   (a) Figure 4.8(a) is a plot of q(t) versus t in two distinct stationary states for two values of q(0) with $\dot{q}(0) = 0$. One starts at q(0) = 2.3, oscillates around $q = 2$, and settles at q = 2 in the steady state; the other one starts at q(0) = 2.5, drops quickly to oscillate around q = 0, and settles at q = 0 in the steady state.
   (b) Figure 4.8(b) is a phase plane plot of $\dot{q}(t)$ versus q(t) for the same initial conditions as (a). As shown, the trajectories spiral into the respective equilibrium point.

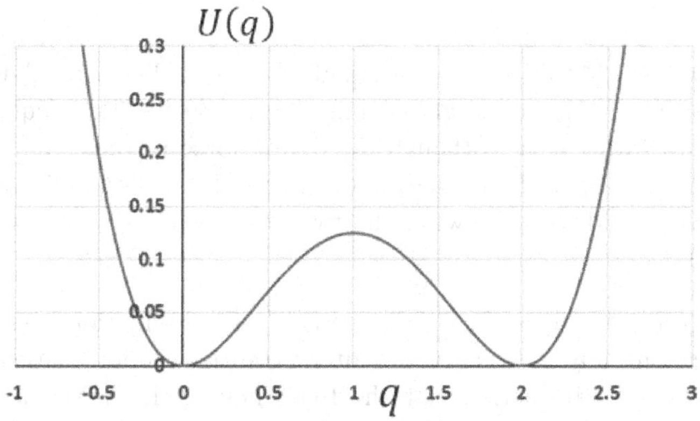

Figure 4.7. Potential of the dynamic system described by Eq. (4.27).

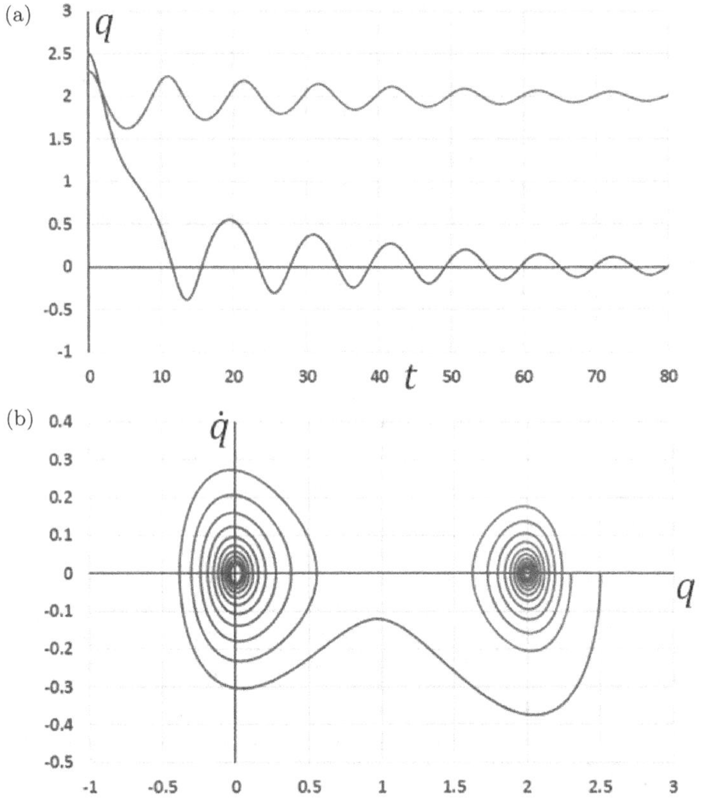

Figure 4.8. (a) Damped nonlinear oscillator responses in the two states and (b) the corresponding phase plane plots.

2. h = 0.05; Eq. (4.27) describes a weakly forced nonlinear oscillator with friction. The oscillator is placed initially at q(0) = 2 (an equilibrium point) with q̇(0) = 0. Although the driven force is weak, the resonant push enables the oscillator to escape the bounded wells to shift the oscillation back and forth between the two wells. This is demonstrated in Fig. 4.9.

   (a) Figure 4.9(a) is a plot of q(t) versus t. The oscillation is not quasi-harmonic anymore. However, after a transient period, the oscillation remains to be periodic at the frequency of the driven force; the driven term of Eq. (4.27) is overlapped to the oscillator response in the plot.

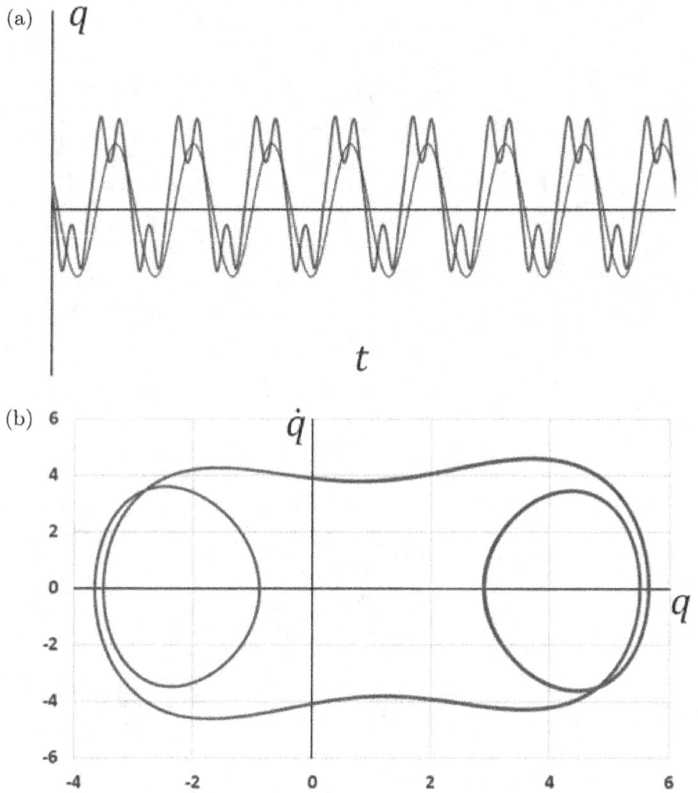

Figure 4.9. (a) Weakly forced nonlinear oscillator response and (b) the corresponding phase plane plot. The driven force term on the RHS of (4.27) is overlapped in (a) for frequency comparison.

(b) Figure 4.9(b) is a phase plane plot of $\dot{q}(t)$ versus $q(t)$ for the same initial conditions as (a). As shown, the trajectory moves back and forth regularly between the two wells.

3. $h = 0.2$; Eq. (4.27) describes a strongly forced nonlinear oscillator with friction. Again, the oscillator is placed initially at $q(0) = 2$ (an equilibrium point) with $\dot{q}(0) = 0$. The resonant push shifts the oscillation back and forth irregularly between the two wells. This is demonstrated in Fig. 4.10.

(a) Figure 4.10(a) is a plot of $q(t)$ versus t. The oscillation has irregular amplitude. Although the oscillation is still bound and have strong

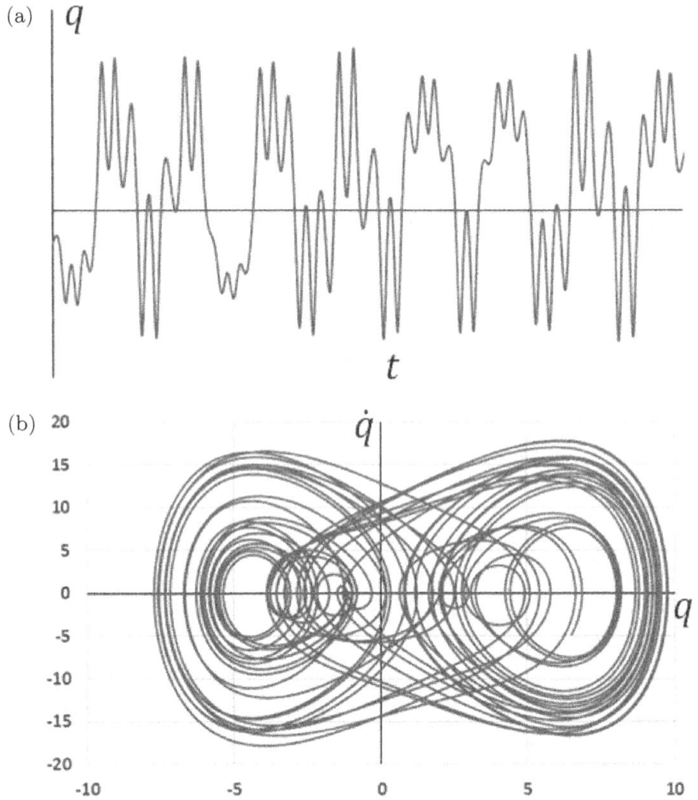

Figure 4.10. (a) Strongly forced nonlinear oscillator response and (b) the corresponding phase plane plot.

spectral components at the driving frequency and its higher harmonics, the oscillation is not periodic (i.e., it doesn't repeat itself from one period to the next). Hence, the oscillator motion becomes unpredictable in detail from one cycle of the driving frequency to the next.

(b) Figure 4.10(b) is a phase plane plot of $\dot{q}(t)$ versus $q(t)$ for the same initial conditions as (a). As shown, the trajectory moves back and forth irregularly between the two wells.

One way to observe the chaotic behavior of the oscillator is to introduce Poincaré section, e.g., phase plane plot with discrete points, which are the trajectory points at each period, i.e., $\left(q\left(t=\frac{2n\pi}{\nu}\right), \dot{q}\left(t=\frac{2n\pi}{\nu}\right)\right)$, where integer $n = 0, 1, 2 \ldots$.

The Poincaré section plots for the three cases are presented in Figs. 4.11(a) to (c). Figure 4.11(a) shows that there are two attractors located at the two stable equilibria. The points mapped on the phase plane show regular patterns. In Fig. 4.11(b), the mapped points show spread, but some regular patterns can still be seen. However, in Fig. 4.11(c), the mapped points are randomly distributed on the phase plane, indicating that the motion of the oscillator becomes chaotic.

4.6 The Van der Pol equation

A nonlinearly stabilized oscillator is defined by

$$\frac{d^2q}{dt^2} - \epsilon(1-\beta q^2)\omega_0\frac{dq}{dt} + \omega_0^2 q = 0 \tag{4.28}$$

where ϵ and β are dimensionless constants with $0 < \epsilon \ll 1$ and $\beta > 0$.

For the initial conditions: $q(0) = q_0$ and $\dot{q}(0) = 0$, the ansatz $q(t) = a(t)\cos\theta$ is chosen, where $d\theta/dt = \Omega(t)$. On substitution into (4.28), with the aid of

$$\frac{dq}{dt} = -\Omega a \sin\theta + \frac{da}{dt}\cos\theta,$$

and

$$\frac{d^2q}{dt^2} = \left(\frac{d^2a}{dt^2} - a\Omega^2\right)\cos\theta - \left(2\frac{da}{dt}\Omega + a\frac{d\Omega}{dt}\right)\sin\theta,$$

defining equations for a and Ω follow from

$$\left[\frac{d^2a}{dt^2} - a(\Omega^2-\omega_0^2) - \epsilon\omega_0\left(1-\frac{3}{4}\beta a^2\right)\frac{da}{dt}\right]\cos\theta$$

$$-\Omega\left[2\frac{da}{dt} + \frac{a}{\Omega}\frac{d\Omega}{dt} - \varepsilon\omega_0 a\left(1-\frac{1}{4}\beta a^2\right)\right]\sin\theta = 0 \tag{4.28a}$$

whence, after setting the coefficients of the $\sin\theta$ and $\cos\theta$ terms equal to zero (since they are independent-the initial phase is arbitrary) and neglecting $\frac{d^2a}{dt^2}$ and $\frac{d\Omega}{dt}$ terms, we find

$$\frac{da}{dt} = \frac{1}{2}\varepsilon\omega_0 a\left(1-\frac{1}{4}\beta a^2\right) \tag{4.28b}$$

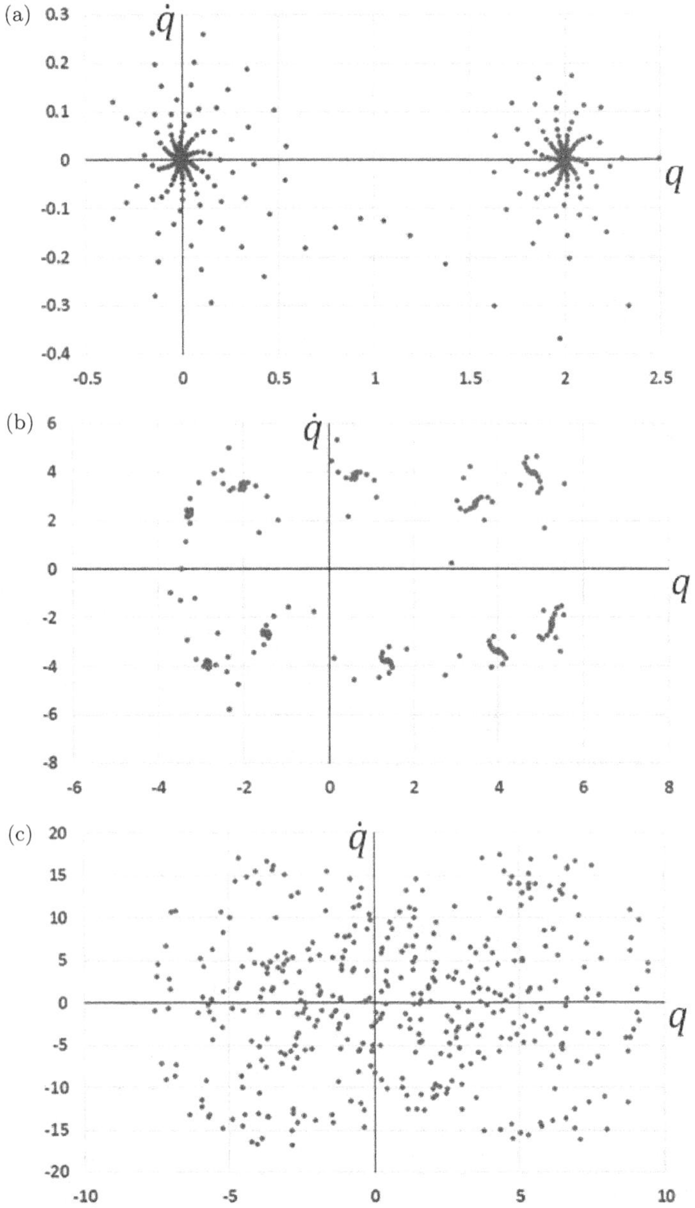

Figure 4.11. Poincaré section plots on the phase plane in the three cases: (a) h = 0; (b) h = 0.05; and (c) h = 0.2. It is showing a progression toward dynamic chaos in responding to the increase of the driven resonance force.

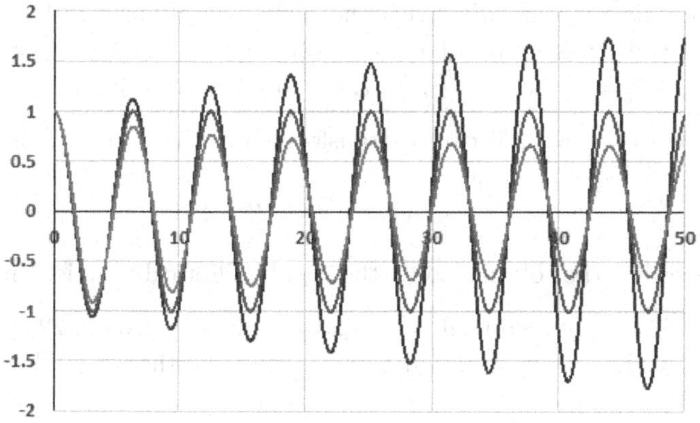

Figure 4.12. Van der Pol oscillator in three cases: 1. Unstable oscillation (blue), 2. Stable oscillation (red), and 4. Damped oscillation (green).

and

$$\Omega^2 = \omega_0^2\left[1 - \frac{\epsilon^2}{2}\left(1 - \frac{1}{4}\beta a^2\right)\left(1 - \frac{3}{4}\beta a^2\right)\right] \tag{4.28c}$$

Equation (4.28b) is integrated to be

$$a(t) = \frac{\frac{2}{\sqrt{\beta}}}{\left[1 + \left(\frac{4}{\beta a_0^2} - 1\right)e^{-\epsilon\omega_0 t}\right]^{\frac{1}{2}}} \tag{4.28d}$$

where $a_0 = a(0) = q_0$.

Based on (4.28c) and (4.28d), solution of (4.28) with $\epsilon = 0.05$ is plotted in Fig. 4.12 for the three cases: $\beta = 1, 4$, and 10, representing unstable, stable, and damped oscillations. These plots match perfectly with the plots from the direct numerical integration.

Problems

P4.1. For an undamped linear oscillator driven by a resonance force as being descriptive by Eq. (4.16) with $\alpha = 0$ and $\nu = \Omega$, Find the analytical solution with the initial conditions: $q(0) = q_0$ and $\dot{q}(0) = 0$.

P4.2. A driven linear oscillator with friction is descriptive by the equation

$$\frac{\mathrm{d}^2 q}{\mathrm{dt}^2} + \gamma\frac{\mathrm{dq}}{dt} + \Omega^2 q = -h\cos\nu\mathrm{t}, \tag{P4.1}$$

Find the response q(t) with the initial conditions: $q(0) = q_0$ and $\dot{q}(0) = 0$ by assuming the ansatz $\mathrm{q(t)} = ae^{-\alpha \mathrm{t}} \cos\theta + \mathrm{b}\cos(\nu \mathrm{t} + \phi)$, where $\frac{\mathrm{d}\theta}{\mathrm{dt}} = \omega$, i.e., $\theta = \theta_0 + \omega t$, a and b are constants.

P4.3. Show that Eq. (4.6) can be transformed to the form:

$$\frac{\mathrm{d}^2 q}{\mathrm{d}\tau^2} + (1 + \mathrm{k}^2)q - 2k^2 q^3 = 0,$$

which has the solution to be the Jacobi elliptical function $\mathrm{sn}(\tau;\ \mathrm{k})$.

P4.4. Consider a damped Duffing oscillator defined by Eq. (4.22), in which $\mathrm{D} > 0$, $\mathrm{E} < 0$, $\mathrm{F} = 0$, and $\mathrm{G} > 0$. Find the three equilibria of the oscillator.

P4.5. Van der Pol oscillator (4.28) has the initial conditions: $\mathrm{q}(0) = 0$ and $\dot{\mathrm{q}}(0) = \mathrm{v}_0$; choose an Ansatz and find the response.

P4.6. Repeat the example considered in Sec. 4.5 with the parameters: $\mathrm{D} = 0.05$, $\mathrm{E} = -(0.2\pi)^2$, $\mathrm{F} = 0$, $\mathrm{G} = (0.2\pi)^2$, and $\nu = 0.2\pi$. Find three representative values of h to show the expansion of the oscillator response from deterministic (regular) to chaotic.

P4.7. A hybrid of the Van der Pol & Rayleigh oscillator defined by the differential equation

$$\ddot{x} - \alpha\dot{x} + \beta\dot{x}^3 + \gamma x^2\dot{x} + \omega^2 x = 0$$

where all coefficients are positive; neglect the higher frequency terms, find the general solution.

Chapter 5

Lagrangian and Hamiltonian Method in One Dimension

5.1 Equations of motion

To obtain a more general procedure for handling nonlinear and time varying oscillatory problems, it is convenient to derive the defining equations for a system described by a single variable $q = q(\mathrm{t})$ in terms of a Lagrangian $L(q, \dot{q}, \mathrm{t})$ which is a functional of q, $\dot{q}$, and t. The defining equation follows from least action principle that

$$\delta \int_{t_1}^{t_2} L(q, \dot{q}, \mathrm{t}) dt = 0 \tag{5.1}$$

for arbitrary δq variations with q fixed at t_1 and t_2, that is,

$$\int_{t_1}^{t_2} \left(\frac{\partial L}{\partial q} \delta q + \frac{\partial L}{\partial \dot{q}} \delta \dot{q} \right) dt = 0.$$

After an integration by parts, and observing that $\delta \dot{q} = \frac{d}{dt}(\delta q)$, the defining oscillatory equation takes the Eulerian form

$$\frac{d}{dt} \left(\frac{\partial L}{\partial \dot{q}} \right) - \frac{\partial L}{\partial q} = 0, \tag{5.2}$$

which is a single higher order differential equation. As an example, for

$$L = \frac{1}{2}(\dot{q}^2 - \Omega^2 q^2 + \alpha^2 q^4),$$

application of (5.2), leads to the source free nonlinear oscillatory problem, namely:

$$\ddot{q} + \Omega^2 q - 2\alpha^2 q^3 = 0. \tag{5.2a}$$

An alternative formulation is obtained by introducing a Hamiltonian function

$$H = \dot{q}\frac{\partial L}{\partial \dot{q}} - L, \tag{5.3}$$

which has the property that (if L is independent of t, i.e., $\frac{\partial L}{\partial t} = 0$ for a stationary system)

$$\frac{dH}{dt} = \ddot{q}\frac{\partial L}{\partial \dot{q}} + \dot{q}\frac{d}{dt}\frac{\partial L}{\partial \dot{q}} - \frac{\partial L}{\partial q}\dot{q} - \frac{\partial L}{\partial \dot{q}}\ddot{q} \equiv 0,$$

whence the Hamiltonian H is invariant with time. Define a momentum variable (conjugate to q) by

$$p = \frac{\partial L}{\partial \dot{q}} = \dot{q},$$

i.e., $H = \dot{q}p - L = H(p, q)$, then on noting that

$$\frac{\partial H}{\partial \dot{q}} = 2p - \frac{\partial L}{\partial \dot{q}} = p,$$

one also obtains from (5.3):

$$\frac{\partial H}{\partial p} = \dot{q} \quad \text{and} \quad \frac{\partial H}{\partial q} = -\frac{\partial L}{\partial q} = -\frac{d}{dt}\frac{\partial L}{\partial \dot{q}} = -\frac{dp}{dt} \tag{5.4}$$

which constitute two first order equations for the determination of q(t) and p(t). Equation (5.4) are the familiar Hamilton's equations. For the example considered above, where

$$L = \frac{1}{2}(\dot{q}^2 - \Omega^2 q^2 + \alpha^2 q^4),$$

hence,

$$p = \frac{\partial L}{\partial \dot{q}} = \dot{q};$$

and the Hamiltonian is

$$H = \frac{1}{2}(p^2 + \Omega^2 q^2 - \alpha^2 q^4),$$

and Hamilton's equations (5.4) become

$$\dot{q} = p \quad \text{and} \quad -\dot{p} = -\ddot{q} = \Omega^2 q - 2\alpha^2 q^3,$$

which reproduce the same nonlinear oscillatory equation (5.2a) as above.

5.2 Average Lagrangian and Hamiltonian method for approximate response

For a system with weakly varying parameters and/or nonlinearity such that

$$q(t) = \begin{cases} a(t)\sin\theta(t) \\ a(t)\cos\theta(t) \end{cases}$$

the rapid oscillatory behavior of $q(\mathrm{t})$ can average out to obtain equations for $a(\mathrm{t})$ and $\omega(\mathrm{t}) = \mathrm{d}\theta/\mathrm{dt}$, which vary only slowly during an oscillation period. The desired equations can be obtained on introduction of an averaged Lagrangian via

$$\mathcal{L} = \frac{1}{2\pi}\int_0^{2\pi} L(q,\dot{q},\mathrm{t})d\theta = \mathcal{L}(a,\omega). \tag{5.5}$$

Note that by writing the averaged Lagrangian as a function of a and ω, we have implicitly neglected terms in $\dot{a} = da/dt$ (a is assumed to be slowly varying); a more accurate procedure, to be presented later, considers the average Lagrangian $\mathcal{L}$ to be a function of $\dot{a}$, as well as a and ω. Instead of the above variational (least action) principle for L relative to arbitrary variations of δq, Whitham has introduced a variational principle for $\mathcal{L}$ that for arbitrary variations of δa and $\delta\theta$ has the form

$$\delta\int_{t_1}^{t_2} \mathcal{L}(a,\omega)dt = 0, \tag{5.6}$$

whence

$$\int_{t_1}^{t_2}\left(\frac{\partial\mathcal{L}}{\partial a}\delta a + \frac{\partial\mathcal{L}}{\partial\omega}\delta\omega\right)dt = 0.$$

Since $\delta\omega = \mathrm{d}(\delta\theta)/\mathrm{dt}$, δa and $\delta\theta$ are arbitrary (except that $\delta\theta(\mathrm{t}_1) = \delta\theta(\mathrm{t}_2) = 0$), one infers a "dispersion relation"

$$\frac{\partial\mathcal{L}}{\partial a} = 0, \tag{5.7a}$$

and a "transport equation"

$$\frac{d}{dt}\frac{\partial\mathcal{L}}{\partial\omega} = 0, \tag{5.7b}$$

which play the role of averaged Euler equations characterizing the macroscopic dynamics of an oscillator in terms of $a(\mathrm{t})$, $\omega(\mathrm{t})$.

5.2.1 *Examples*

Reconsidering in Lagrangian terms the similar cases as those discussed in Sec. 4.3 of Chapter 4, but source-free, i.e., h = 0. The source-free equation (5.2a), with prescribed initial conditions $q(0) = 0$ and $\dot{q}(0) = \mathrm{v}_0$, is analyzed for

(a) linear: $\Omega = \text{constant}$, $\alpha = 0$
(b) linear: $\Omega = \Omega(\mathrm{t})$, $\alpha = 0$
(c) nonlinear: $\Omega = \Omega(\mathrm{t})$, $\alpha \neq 0$

One finds in case (a) and (b) that

$$L = \frac{1}{2}(\dot{q}^2 - \Omega^2 q^2), \quad \text{then} \quad \mathcal{L} = \frac{1}{4}(a^2\omega^2 - \Omega^2 a^2),$$

and hence, the average dispersion relation follows from (5.7a) as:

$$\omega^2 - \Omega^2 = 0, \quad \text{that is } \omega = \pm\Omega(t). \tag{5.8a}$$

The transport equation (5.7b) implies that $a^2\omega$ is an adiabatic invariant, namely:

$$\frac{d}{dt}\left(\frac{a^2\omega}{2}\right) = 0. \tag{5.8b}$$

Hence, for the initial conditions $q(0) = 0$, $\dot{q}(0) = \mathrm{v}_0 \geq 0$,

$$q(t) = a(t)\sin\theta(t) = a_0\sqrt{\frac{\Omega_0}{\Omega}}\sin\left(\int_0^t \Omega(t')dt'\right) \tag{5.8c}$$

where $\Omega_0 = \Omega(0)$ and $a_0 = a(0) = \mathrm{v}_0/\Omega_0$.

Similarly, for case (c), where

$$L = \frac{1}{2}(\dot{q}^2 - \Omega^2 q^2 + \alpha^2 q^4),$$

then

$$\mathcal{L} = \frac{1}{4}\left(a^2\omega^2 - \Omega^2 a^2 + \frac{3}{4}\alpha^2 a^4\right),$$

and the dispersion relation follows as

$$\frac{\partial \mathcal{L}}{\partial a} = \frac{1}{2}\left(\omega^2 a - \Omega^2 a + \frac{3}{2}\alpha^2 a^3\right) = 0,$$

from which one finds

$$\omega = \pm\sqrt{\Omega^2 - \frac{3}{2}\alpha^2 a^2}. \tag{5.8d}$$

As before, the transport equation yields

$$\frac{d}{dt}\frac{\partial \mathcal{L}}{\partial \omega} = \frac{d}{dt}\left(\frac{a^2\omega}{2}\right) = 0, \tag{5.8e}$$

from which $a^2\omega$ is again an adiabatic invariant, however, $\omega \neq \Omega$.

Hence,

$$q(t) = a(t)\sin\theta(t) = a_0\sqrt{\frac{\omega_0}{\omega}}\sin\left(\int_0^t \omega(t')dt'\right), \tag{5.8f}$$

where $\omega_0 = \omega(0) = \alpha\sqrt{\Omega_0^2 - \frac{3}{2}\alpha^2 \mathrm{a}_0^2}$ and $a_0 = a(0) = \mathrm{v}_0/\omega_0$.

Alternatively, one can introduce an averaged Hamiltonian via

$$\mathcal{H} = \frac{1}{2\pi}\int_0^{2\pi} H\,d\theta. \tag{5.9}$$

From which for cases (a) and (b)

$$\mathcal{H} = \frac{1}{2\pi}\int_0^{2\pi} \frac{1}{2}(\dot{q}^2 + \Omega^2 q^2)\,d\theta = \frac{1}{4}(\omega^2 + \Omega^2)a^2 = \frac{\Omega^2 a^2}{2},$$

whence $\frac{d}{dt}(\frac{\mathcal{H}}{\omega}) = 0$ and $\frac{\mathcal{H}}{\omega}$ is recognized as an adiabatic invariant. Describe the oscillation in terms of an aggregation of quanta (quasi-particles), and the energy of each quasi-particle is proportional to ω, hence, $\frac{\mathcal{H}}{\omega}$ is proportional to the number of quanta in the system. It shows that the number of quanta is conserved. However, for case (c)

$$\begin{aligned}\mathcal{H} &= \frac{1}{2\pi}\int_0^{2\pi} \frac{1}{2}(\dot{q}^2 + \Omega^2 q^2 - \alpha^2 q^4)d\theta \\ &= \frac{1}{4}\left(\omega^2 + \Omega^2 - \frac{3}{4}\alpha^2 \mathrm{a}^2\right)a^2\end{aligned}$$

and one sees that

$$\frac{d}{dt}\left(\frac{\mathcal{H}}{\omega}\right) \neq 0. \tag{5.9a}$$

It is realized because those quanta generated at harmonic frequencies have been neglected in the average procedure; in other words, $\frac{\mathcal{H}}{\omega}$ does not represent the total number of quanta in the system.

A more accurate procedure is obtained by retaining the terms in $\dot{a}$ that were neglected in the preceding analysis. For case (c) the averaged

Lagrangian is then a function of a, $\dot{a}$, and ω as follows:

$$\mathcal{L}(a, \dot{a}, \omega) = \frac{1}{2\pi}\int_0^{2\pi} L d\theta = \frac{1}{4}\left(a^2\omega^2 - \Omega^2 a^2 + \dot{a}^2 + \frac{3}{4}\alpha^2 \mathrm{a}^4\right).$$

The Whitham variational principle when applied to $\mathcal{L}(a, \dot{a}, \omega)$ yields the "dispersion relation"

$$\frac{\partial \mathcal{L}}{\partial a} - \frac{d}{dt}\frac{\partial \mathcal{L}}{\partial \dot{a}} = 0, \tag{5.10}$$

From which we obtain

$$\omega^2 = \Omega^2 - \frac{3}{2}\alpha^2 \mathrm{a}^2 + \frac{\ddot{a}}{a}. \tag{5.11}$$

Equation (5.10) is evidently a generalization of (5.7a). The transport equation following from the Whitham variational principle in this case is identical to that previously obtained in (5.7b); thus, once again we find that $a^2\omega$ is an adiabatic invariant. However, this more accurate procedure still cannot preserve $\frac{\mathcal{H}}{\omega}$ adiabatic invariant. When $\ddot{a} \ll a$, the more accurate dispersion relation (5.11) reduces to the result (5.8) obtained previously from the simpler averaging method.

5.3 Averaging for strongly nonlinear variable parameter systems

For variable parameter nonlinear systems, one can no longer use the quasi-harmonic approximation for q because those quanta generated at harmonic frequencies have been neglected in the average procedure, evidenced by (5.9a). In the following analysis, q is set to be a more general periodic function of θ (i.e. $q = q(\theta)$ — not the same function as $q(\mathrm{t})$) of period 2π, but the variations of the system parameters in one period is restricted to be small; a likely periodic function is one of the Jacobi elliptic functions. According to the same procedure as before, one introduces the averaged Lagrangian $\mathcal{L}_s$ by

$$\mathcal{L}_s = \frac{1}{2\pi}\int_0^{2\pi} L(q, \dot{q})\, d\theta, \tag{5.12}$$

with

$$\mathrm{q} = \mathrm{q}(\theta), \quad \omega = \frac{d\theta}{dt}, \quad \text{and} \quad \dot{q} = \omega\frac{dq}{d\theta}.$$

Using the relation (5.3), one rewrites (5.12) as

$$\mathcal{L}_s = \frac{1}{2\pi}\int_0^{2\pi} [p\dot{q} - H]\, d\theta = \frac{1}{2\pi}\int_0^{2\pi} \omega p\, dq - \mathcal{H}_s, \tag{5.13a}$$

where

$$\mathcal{H}_s = \frac{1}{2\pi}\int_0^{2\pi} H\, d\theta, \tag{5.13b}$$

and the integration is over a complete cycle of q.

Make the adiabatic assumption that ω and H do not vary appreciably over one period in θ, leads to the approximations

$$\int \omega p\, dq \cong \omega \int p\, dq \quad \text{and} \quad \mathcal{H}_s \cong H.$$

Since $\mathrm{H} = \mathrm{H(p, q)} \cong \mathcal{H}_s$, one infers that $\mathrm{p} = \mathrm{p}(q, \mathcal{H}_s)$, whence the "action" integral

$$S = \frac{1}{2\pi}\int p dq = \frac{2}{\pi}\int_0^{q_{max}} \dot{q} dq = S(\mathcal{H}_s), \tag{5.14}$$

is a function only of $\mathcal{H}_s$, where q_{max} is the maximum of q(t) and

$$p = \dot{q} = \pm\sqrt{2(H - V(q))}.$$

Thus, the averaged Lagrangian becomes

$$\mathcal{L}_s = \omega S(\mathcal{H}_s) - \mathcal{H}_s = \mathcal{L}_s(\mathcal{H}_s, \omega), \tag{5.15}$$

and is a function of the adiabatic quantities $\omega, \mathcal{H}_s$ — which differs from the previous analysis of averaging in that the amplitude a (which is difficult to define in the nonlinear case) is used to obtain $\mathcal{L}(a, \omega)$.

Noting that ω and $\mathcal{H}_s$ are weak functions of the variable t, with $\omega = \mathrm{d}\theta/\mathrm{dt}$; thus, the stationarity principle is applied

$$\delta \int_{t_1}^{t_2} L(q, \dot{q}) dt = 0 = \int_{t_1}^{t_2} \left(\frac{\partial \mathcal{L}_s}{\partial \mathcal{H}_s}\delta\mathcal{H}_s + \frac{\partial \mathcal{L}_s}{\partial \omega}\delta\omega \right) dt,$$

with $\delta\omega = \frac{d}{dt}\delta\theta$; it becomes after an integration by parts

$$\int_{t_1}^{t_2} \left[\frac{\partial \mathcal{L}_s}{\partial \mathcal{H}_s}\delta\mathcal{H} - \frac{d}{dt}\left(\frac{\partial \mathcal{L}_s}{\partial \omega}\right)\delta\theta \right] dt = 0.$$

The averaged Eulerian equations are then obtained to be

$$\frac{\partial \mathcal{L}_s}{\partial \mathcal{H}_s} = 0 \quad \text{and} \quad \frac{d}{dt}\frac{\partial \mathcal{L}_s}{\partial \omega} = 0, \tag{5.16a}$$

or

$$\omega\frac{\partial S}{\partial \mathcal{H}_s} = 1 \quad \text{and} \quad \frac{dS}{dt} = 0. \tag{5.16b}$$

Thus, one infers the "dispersion relation"

$$\omega = 1/(\partial S/\partial \mathcal{H}_s),$$

and the fact that $S(\mathcal{H}_s)$ is an adiabatic invariant.

Note that $\mathcal{H}_s \cong H(p,q)$, whence

$$p = p(\mathcal{H}_s, q), \quad \text{and} \quad \frac{\partial S}{\partial \mathcal{H}_s} = \frac{1}{2\pi}\int \frac{\partial p}{\partial \mathcal{H}_s} dq;$$

the quantity $\frac{\partial S}{\partial \mathcal{H}_s}$ is sometimes easier to evaluate than S.

As an example, the same equation (5.2a) is considered; thus,

$$p = \dot{q} = \pm\sqrt{2H - \Omega^2\left(1 - \frac{q^2}{2q_M^2}\right)q^2} \cong \pm\sqrt{2\mathcal{H}_s - \Omega^2\left(1 - \frac{q^2}{2q_M^2}\right)q^2},$$

where $q_M = \frac{\Omega}{\sqrt{2}\alpha}$; and then

$$\frac{\partial S}{\partial \mathcal{H}_s} = \frac{2}{\pi\alpha}\int_0^{q_{2s}} \frac{dq}{\sqrt{(q^2 - q_{1s}^2)(q^2 - q_{2s}^2)}} = \frac{2}{\pi\alpha}\frac{K(\beta_s)}{q_{1s}}$$

where

$$q_{1s}^2 = q_M^2\left(1 + \sqrt{1 - \frac{4\mathcal{H}_s}{\Omega^2 q_M^2}}\right) \tag{5.16c}$$

$$q_{2s}^2 = q_M^2\left(1 - \sqrt{1 - \frac{4\mathcal{H}_s}{\Omega^2 q_M^2}}\right) \tag{5.16d}$$

$$K(\beta_s) = \int_0^1 \frac{d\xi}{\sqrt{(1-\xi^2)(1-\beta_s^2\xi^2)}} = sn^{-1}(1, \beta_s^2) \tag{5.16e}$$

and $\beta_s^2 = \frac{q_{2s}^2}{q_{1s}^2}$. Hence,

$$\omega = \frac{\pi\alpha q_{1s}}{2K(\beta_s)}. \tag{5.16f}$$

With the prescribed initial conditions $q(0) = 0$ and $\dot{q}(0) = \mathrm{v}_0$, $2\mathcal{H}_s = \mathrm{v}_0^2$; then

$$q(t) = q_2\,\mathrm{sn}\left(\frac{2K(\beta_s)}{\pi}\int_0^{\mathrm{t}} \omega(\mathrm{t}')\mathrm{dt}'\right), \tag{5.16g}$$

where

$$q_2 = \frac{\Omega}{\sqrt{2}\alpha}\left(1 - \sqrt{1 - \frac{4\alpha \mathrm{v}_0^2}{\Omega^4}}\right)^{1/2}.$$

5.4 Analytical approach for strongly nonlinear variable parameter lumped systems

For systems which have periodic wave responses to excitation, the techniques of analysis considered up to now have been restricted to weakly variable and nonlinear systems, whose response can be regarded as close to the stationary response of a linear system that is of the form

$$q(t) = a(t)\begin{Bmatrix} \cos\theta(t) \\ \sin\theta(t) \end{Bmatrix}, \quad \text{with } \frac{d\theta}{dt} = \omega,$$

where $a(\mathrm{t})$ and $\omega(\mathrm{t})$ are assumed weakly dependent on the one dimensional variable t. "Weak" implies that the variability of a and ω is small in a period $2\pi/\omega$ of the system response, and that nonlinear terms are small compared to the linear terms (i.e., amplitude a is small). Thus, the response of such systems is of the quasi-harmonic form. One should note that, whereas the Fourier spectrum of a constant parameter, linear system is a point spectrum, the Fourier spectrum of an excited, weakly nonlinear and variable parameter system is not of this simple type; in the weak case, the spectrum contains only a small number of significant points.

For strongly nonlinear, variable parameter systems, the analysis can no longer be assumed to be of a quasi-harmonic character; a more general average approach assuming the ansatz: $q = q(\theta)$, a more general periodic function of θ, is presented in Sec. 5.3. But it is necessary to examine the accuracy of the average methods. The justification procedure is feasible, because the "stationary" responses of certain constant parameter linear and nonlinear systems have analytical forms; those results can be based to check the accuracy of the approximate responses obtained from the average methods. The comparisons are presented in the following.

If we characterize such a nonlinear system by a Lagrangian $L(q, \dot{q}, \mathrm{t})$, or a Hamiltonian $H(p, q, t)$, then as before, its equation of motion (i.e. the nonlinear equation of interest) is defined by the Eulerian equation

$$\frac{d}{dt}\left(\frac{\partial L}{\partial \dot{q}}\right) - \frac{\partial L}{\partial q} = 0,$$

or by Hamilton 's equations

$$\frac{dq}{dt} = \frac{\partial H}{\partial p} \quad \text{and} \quad \frac{dp}{dt} = -\frac{\partial H}{\partial q}.$$

For the problem considered in Sec. 5.2, the stationary response with Ω, α constant is deducible from a Lagrangian or Hamiltonian of the form

$$L = \frac{1}{2}(\dot{q}^2 - \Omega^2 q^2 + \alpha^2 q^4), \tag{5.17a}$$

or

$$H = \frac{1}{2}(\dot{q}^2 + \Omega^2 q^2 - \alpha^2 q^4), \tag{5.17b}$$

which evidently characterizes the dynamics of an equivalent unit mass particle of position $q(t)$ in a potential field. The equivalent kinetic and potential energies are identified as

$$T(\dot{q}) = \frac{\dot{q}^2}{2} \quad \text{and} \quad V(q) = \frac{1}{2}(\Omega^2 q^2 - \alpha^2 q^4). \tag{5.18}$$

For the stationary case Ω = constant, α = constant, the solution q(t) can readily be pictured in terms of a potential versus q plot as shown in Fig. 5.1, in which $q_M = \frac{\Omega}{\sqrt{2}\alpha}$. The solution q(t), for potentials of the indicated form, can be obtained analytically. In the following, derivation of the exact function representations of $q(t)$ is first presented.

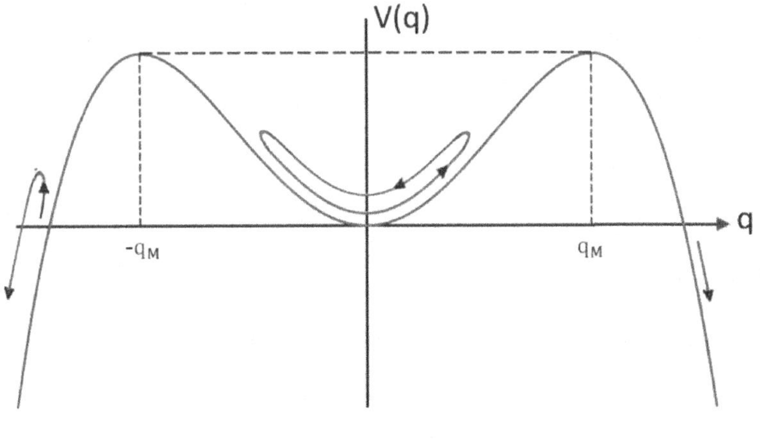

Figure 5.1. Potential plot of equivalent particle dynamic problem.

The relationship $T(\dot{q}) = H - V(q)$ leads to

$$\dot{q} = \sqrt{2(H - V(q))} = \sqrt{2H - \Omega^2\left(1 - \frac{q^2}{2q_M^2}\right)q^2},$$
$$= \frac{\Omega}{\sqrt{2}q_M}\sqrt{(q^2 - q_1^2)(q^2 - q_2^2)}, \tag{5.19}$$

where

$$q_1^2 = q_M^2\left(1 + \sqrt{1 - \frac{4H}{\Omega^2 q_M^2}}\right) \tag{5.20a}$$

and

$$q_2^2 = q_M^2\left(1 - \sqrt{1 - \frac{4H}{\Omega^2 q_M^2}}\right) \tag{5.20b}$$

whence

$$\int_0^q \frac{dq}{\sqrt{(q^2 - q_1^2)(q^2 - q_2^2)}} = \frac{\Omega}{\sqrt{2}q_M}t, \tag{5.21a}$$

if q(0) = 0 and $H < \Omega^2 q_M^2/4$. With the substitutions

$$\xi^2 = \frac{q^2}{q_2^2} \quad \text{and} \quad \beta^2 = \frac{q_2^2}{q_1^2}$$

We find

$$\int_0^{q/q_2} \frac{d\xi}{\sqrt{(1-\xi^2)(1-\beta^2\xi^2)}} = \frac{q_1}{\sqrt{2}q_M}\Omega t$$
$$= sn^{-1}\left(\frac{q}{q_2}, \beta^2\right). \tag{5.21b}$$

The integral can be expressed in terms of a Jacobi elliptic function as indicated. The resulting expression for q is

$$q = q_2\, sn\left(\frac{q_1}{\sqrt{2}q_M}\Omega t, \beta^2\right).$$

Since q oscillates between $\pm q_2$, a complete cycle of q corresponds to a period in t, which yields a dispersion relation for ω as follows:

$$4K(\beta) = \frac{q_1\Omega}{\sqrt{2}q_M\omega}2\pi$$

where the complete elliptic integral of the first kind is defined as

$$K(\beta) = \int_0^1 \frac{d\xi}{\sqrt{(1-\xi^2)(1-\beta^2\xi^2)}}$$

Thus, we find

$$\omega = \frac{q_1\Omega}{q_M \frac{2\sqrt{2}}{\pi}K(\beta)} = \frac{\Omega}{\sqrt{2}} \frac{\left[1+\sqrt{1-\frac{8H\alpha^2}{\Omega^4}}\right]^{1/2}}{\frac{2}{\pi}K(\beta)}, \tag{5.21c}$$

which is the same as (5.16f) when $\mathcal{H}_s = H$ is assumed.

Hence, the stationary response is

$$q = q_2\, sn\left[\frac{2}{\pi}K(\beta)\omega t\right], \tag{5.22a}$$

It is the same as (5.16g) with constant ω.

In the case of small β (that is, $\frac{4H}{\Omega^2 q_M^2} \ll 1$)

$$\frac{2}{\pi}K(\beta) \cong 1 + \frac{\beta^2}{4} + \cdots$$

Moreover,

$$\begin{aligned} sn\left[\frac{2}{\pi}K(\beta)\omega t\right] &\cong \left(1+\frac{\beta^2}{4}\cos^2\theta\right)\sin\theta + \cdots \\ &\cong \left(1+\frac{\beta^2}{16}\right)\sin\theta + \frac{\beta^2}{16}\sin 3\theta + \cdots \end{aligned}$$

and

$$\beta^2 = \frac{q_2^2}{q_1^2} = \frac{1-\sqrt{1-\frac{4H}{\Omega^2 q_M^2}}}{1+\sqrt{1-\frac{4H}{\Omega^2 q_M^2}}} \cong \frac{H}{\Omega^2 q_M^2}.$$

Equation (5.22a), to order β^2, reduces to

$$q = \sqrt{2}q_M\beta\left(1+\frac{\beta^2}{2}\right)\left[\left(1+\frac{\beta^2}{16}\right)\sin\theta + \frac{\beta^2}{16}\sin 3\theta + \cdots\right] \tag{5.22b}$$

and

$$\omega \cong \frac{\Omega}{1+\frac{3}{4}\beta^2} = \frac{\Omega}{1+\frac{3}{4}\frac{H}{\Omega^2 q_M^2}} \cong \Omega\sqrt{1-3\frac{H\alpha^2}{\Omega^4}}, \tag{5.22c}$$

which agrees with the result previously obtained in the quasi-linear case. The linear limit $\beta \cong 0$ is evidently

$$q = \frac{\sqrt{2H}}{\Omega}\sin\theta \quad \text{and} \quad \omega = \Omega.$$

In the case of $H = \Omega^2 q_M^2/4$, $q_1 = q_M = q_2$ and $\beta = 1$, hence,

$$q = q_M \, sn\left(\frac{\Omega}{\sqrt{2}}t, 1\right) = q_M \tanh \frac{\Omega}{\sqrt{2}}t. \tag{5.22d}$$

The exact response (5.22a) is shown to be the same as (5.16g) with constant ω, evidencing that the extended average method, based on Whitham variational principle, is applicable in the strong nonlinear regime.

Since the energy H is a constant of the motion ($dH/dt = \partial H/\partial t = 0$), its value at $t = 0$ determines the energy for all t. The intuitive knowledge of dynamics in a potential shown in Fig. 5.1 and with the aid of the analytical representations in (5.22a) and (5.22d), suggest that the nature of the q(t) dependence on t can be identified in plots for different levels of H value. Thus, if we consider the stationary response defined by $q(0) = 0$, $\dot{q}(0) = v_0 \geq 0$ prescribed, where $H = \dot{q}^2(0)/2 = v_0^2/2$, then, three cases must be considered depending on the magnitude of H:

(a) $H < \Omega^2 q_M^2/4 = V_{max}$, where $q_M = \frac{\Omega}{\sqrt{2}\alpha}$; $q(t)$ oscillates in the potential well with $-q_M < q < q_M$ (stable case).
(b) $H = \Omega^2 q_M^2/4 = V_{max}$, $q(t)$ increases with time to approach q_M, as the equivalent particle is climbing the potential to stop at the peak of the potential at $q = q_M$.
(c) $H > \Omega^2 q_M^2/4 = V_{max}$, $q(t)$ overshoots the potential maximum and increases monotonically with t (unstable case).

Case (a) It leads to a stationary periodic wave response for $q(t)$, which has an analytical function

$$q(t) = q_2 \, \mathrm{sn}\left(\frac{q_1}{\sqrt{2}q_M}\Omega t, \beta^2\right),$$

where

$$q_1 = q_M \left(1 + \sqrt{1 - \frac{4H}{\Omega^2 q_M^2}}\right)^{\frac{1}{2}},$$

$$q_2 = q_M \left(1 - \sqrt{1 - \frac{4H}{\Omega^2 q_M^2}}\right)^{\frac{1}{2}}, \quad \text{and} \quad \beta = \frac{q_2}{q_1}.$$

An example with $\Omega^2 = 1$, $\alpha^2 = 1$, and $H = 0.05$, the analytical function q(t) is plotted in Fig. 5.2(a) as Plot A; the approximate function (5.8f) is also plotted as Plot B, which is overlapped to plot A for a close comparison.

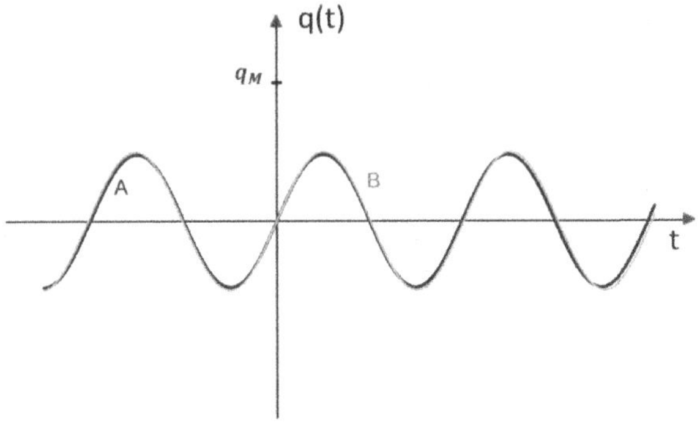

(a) $H << V_{max}$.

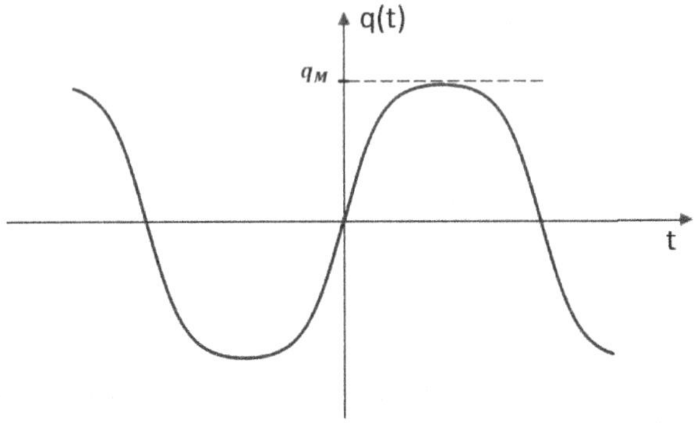

(b) $H \cong V_{max}$, but $H < V_{max}$.

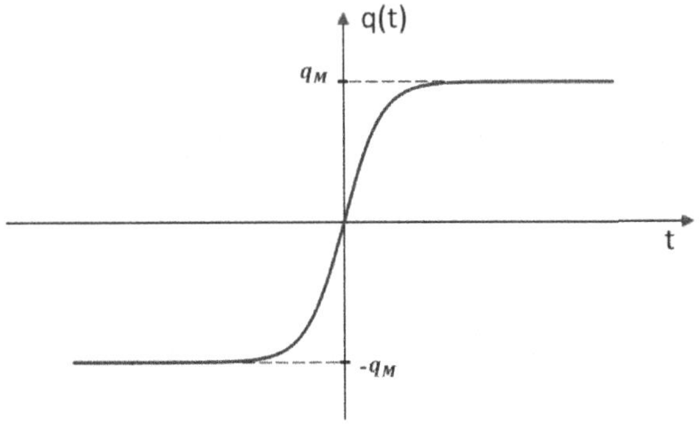

(c) $H = V_{max}$.

Figure 5.2. Nonlinear oscillator response.

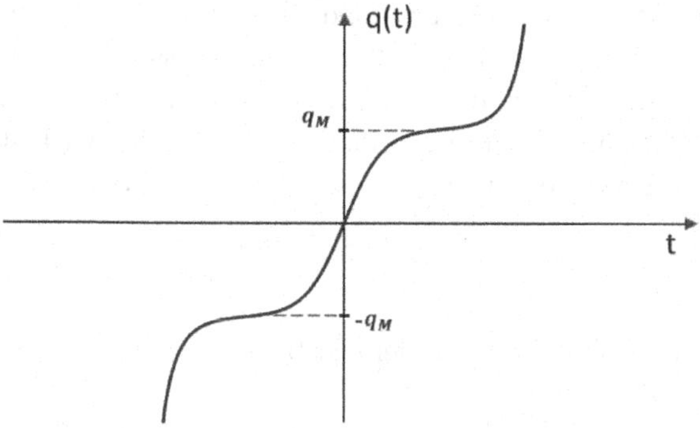

(d) $H > V_{max}$.

Figure 5.2. (*Continued*)

As shown, the average Lagrangian method yields a very good approximate result.

As $H > V_{max}/2$, the approximate response in the function form of (5.8f) starts to show deviation. For strongly nonlinear systems, the stationary response with H close to V_{max} is a nonharmonic function; but is still periodic. For example, as H is increased to 0.12475, which is very close to $V_{max} = 0.125$, the periodic response, as shown in Fig. 5.2(b), deforms significantly from a sinusoidal function. Hence, (5.8f) is not an appropriate approximate response anymore; on the other hand, the approximate response (5.16g) (turn out to be the exact response) has exact match.

Case (b) $H = V_{max}$ leads to an aperiodic (bistatic) response

$$q(t) = q_M \tanh \frac{\Omega}{\sqrt{2}} t,$$

as shown in Fig. 5.2(c), because the particle remains at the potential peak for an infinitely long time; but it should be noted that the latter location is not stable. There is a different potential shape if α^2 is negative!

Case (c) $H > V_{max}$ leads to an unstable response.

After the equivalent particle climbs over the potential maximum, it is accelerated continuously by the background force and increases monotonically with t, as shown in Fig. 5.2(d).

In sum, an averaging technique presented in Sec. 5.2 can be performed for variable parameter systems, and will lead to a good approximation for the amplitude and frequency, provided $H < V_{max}/2$ and the parameter variability is weak — that is, small variation over a period of the wave. This average Lagrangian and Hamiltonian method is further developed in Sec. 5.3, via Whitham's variational method, for dealing with strongly non-linear variable parameter systems.

Problems

P5.1. A one-dimensional system defined by

$$\frac{d^2q}{dt^2} + Eq + Fq^2 + Gq^3 = 0$$

(1) Define the Lagrangian $L(q, \dot{q}, \mathrm{t})$ of the system
(2) Choose an Ansatz of q(t) to determine the average Lagrangian $\mathcal{L}(a, \omega)$
(3) Apply the dispersion relation and transport equation to obtain the approximate response.

P5.2. Reconsider P2.3.; the plasma density distribution in the bottomside of the ionosphere is modelled by an Epstein profile to be $\mathrm{n(z)} = \mathrm{n_0}(1 - \mathrm{e}^{-\frac{z}{L}})$ where $\mathrm{z} = 0$ is located at the lower boundary of the ionosphere. Plasma has a dispersion $\omega = \sqrt{\omega_\mathrm{p}^2 + \mathrm{k^2c^2}}$, where $\omega_\mathrm{p}^2 = \omega_0^2(1 - \mathrm{e}^{-\frac{z}{L}})$ and $\omega_0^2 = \frac{\mathrm{n_0 e^2}}{\mathrm{m_e}\epsilon_0}$. A wave at frequency ω_0 is transmitted upward vertically from the ground into the ionosphere.

(1) Show the Helmholtz equation of the wave field $E(\mathrm{z})$ in the ionosphere
(2) Define an equivalent Lagrangian $L(E, \dot{E}, \mathrm{z})$ of the system, where $\dot{E} = \frac{dE}{dz}$
(3) Apply average Lagrangian method to obtain approximate response
(4) Compare the result with the WKB solution.

P5.3. Apply the equivalent Lagrangian $L(E, \dot{E}, \mathrm{z})$ of the system, obtained for the question (2) of P5.2,

(1) Define an equivalent Hamiltonian $\mathrm{H}(E, \dot{E}, \mathrm{z})$ of the system
(2) Show that the Hamilton's equations of motion are combined to a Helmholtz equation, which is the same as that shown in (1) of the problem P5.2.

P5.4. An oscillator with a position-dependent effective mass $m = 1 + \lambda x^2$ is described by a Lagrangian

$$L = \frac{1}{2(1+\lambda x^2)}(\dot{x}^2 - \alpha^2 x^2)$$

(1) Define the equation of the oscillator
(2) Assume the ansatz: $x(t) = A\cos(\omega t + \varphi)$, find the solution and show that $\omega = \frac{\alpha}{\sqrt{1+\lambda A^2}}$.

P5.5. A system is descriptive by a nonlinear equation

$$\ddot{x} + 3kx\dot{x} + k^2x^3 = 0$$

(1) Determine the Lagrangian $L(x, \dot{x}, t)$ of the system
(2) Find the solution of the equation

P5.6. A nonlinear oscillator is descriptive by the Lagrangian

$$L = \frac{1}{k\dot{x} + k^2x^2 + \omega^2}$$

(1) Define the governing equation of the oscillator
(2) Introduce $x = \frac{1}{k}\frac{\dot{y}}{y}$ (the Cole-Hopf transformation), show that the nonlinear oscillator is transformed to a linear one
(3) Find the general solution of the nonlinear differential equation

P5.7. With the aid of Eq. (5.16b), verify the result of Eq. (5.16g).

Chapter 6

Nonlinear Waves

6.1 Introduction

Nonlinear wave phenomena occur in the propagation of waves of high and low frequencies through media, where the wave dependent changes in the properties of media become significant. The solutions of the nonlinear wave equations often exhibit unique phenomena, such as stable localized waves (e.g., solitons), self-similar structures, chaotic dynamics and wave discontinuities such as shock waves (derivative singularities), and/or wave collapse (where the solution tends to infinity in finite time or finite propagation distance). Nonlinear waves are of wide physical and mathematical interest and practical applications, in a variety of areas such as nonlinear optics, fluid dynamics, plasma physics, etc.

A distinction between linear systems and nonlinear systems is that the superposition principle is not applicable for nonlinear systems. A wave perturbation in a linear system can be obtained via a linear combination of the eigen wave modes, which follow the linear wave dispersion relations of the system and do not interact each other. On the other hand, it is not possible to predict the evolution of the nonlinear wave based on the characteristics of these linear eigen modes because these modes interact one another. Moreover, nonlinear media do not, in general, admit constant speed propagation of waves with arbitrary amplitude and shape. There exist different categories of nonlinear media; for certain amplitudes, some admit the propagation of (periodic or pulse) waves of definite shape with a constant speed; in others, the admitted waves have neither a definite shape nor a constant speed. Waves that can be propagated with constant speed and shape are called stationary waves. On the other hand, nonlinear waves have neither a constant speed nor shape-that is, they are "non-stationary". There is also a special class of such waves called "simple waves" that are quasi-stationary in certain ranges.

In linear systems, a wave impingent on a discontinuity (spatial or temporal) in the medium gives rise to a multiplicity of scattered waves. The scattered waves are characteristic of the various "channels" associated with the discontinuity. The determination of the scattered waves, their amplitudes, form, etc., is well developed for linear systems.

In nonlinear systems there occur, in addition to space-time discontinuities, "self-induced discontinuity" regions wherein the amplitude or phase of the wave field changes very rapidly, or where the amplitude suddenly becomes very large. In either event there arises a multiplicity of scattered waves. One has to distinguish, as in the linear case, between the waves or "mode types" that can exist in the various "channels" associated with the discontinuity or transition region. For example, an incident wave on such a discontinuity may be a nonstationary nonlinear wave that steepens and breaks at the discontinuity, then on leaving the transition region, it decomposes into different types of stationary waves. The determination of the wave behavior in the transition region at the discontinuity is a major challenge. A general wave description near a discontinuity is complicated by the presence of a number of wave types, i.e., the differential equations are of high order. The reduction of these equations to a number of generic first order equations, to be presented in Sec. 6.2, descriptive of different wave types is a necessary first step. These lead to the methods of characteristic curves and Riemann invariants in some cases.

The determination of the possible stationary waves, and more generally of nonstationary waves, in a given nonlinear medium is a basic kinematic concern. The analytical technique for their determination is dependent upon the nature of these waves, for example, whether they are periodic, aperiodic, or quasi-periodic. Periodic or aperiodic waves of stationary type may be sought by solving the space-time, (x, t) dependent partial differential wave equation in a constantly moving wave frame, defined by a coordinate $\xi = \mathrm{x} - \mathcal{U}\mathrm{t}$ or $\xi = \mathrm{kx} - \omega\mathrm{t}$, traveling with the wave. In terms of ξ the wave differential equations are ordinary, at least in one dimension, and may be solved by quadrature in many cases. For "simple" nonstationary waves, one employs a local wave frame characterized by the speed $\mathcal{U}(\xi)$ and seeks solutions dependent only on ξ. In general, the response to excitation gives rise to nonstationary waves, and in one dimension x, requires solving partial differential equations in both $\xi = x - \mathcal{U}t$, and $\eta = x$, where $\mathcal{U}$ may be chosen as the wave velocity for waves of small amplitude. For such quasi-periodic waves, one has available solution techniques which are related to the method of adiabatic invariants, "asymptotic" theories, Whitham's variational method, and related "averaging" procedures presented in Chapter 5.

Stationary solutions, if they exist, constitute the "mode" types of a nonlinear system. In a suitably moving frame, all are wave solutions that have an invariant form, namely: $f(x,t) = f(\xi)$. Stationary solutions may be periodic or aperiodic. In the linear limit when the wave amplitude approaches zero, periodic waves become sinusoidal; solitons and shocks no longer are present (i.e., they degenerate to a trivial "constant" type solution). Note that the wave structure (amplitude, velocity, etc.) is constant in space and time for a stationary solution. Examples of both types (periodic and aperiodic) of stationary nonlinear waves are displayed in the following:

(1) For low frequency or baseband wave phenomena in nonlinear systems, stationary solutions take the forms shown in Figs. 6.1(a)–(e).
(2) For high frequency or carrier wave phenomena in nonlinear systems, stationary solutions take the forms shown in Figs. 6.2(a)–(c).

In the presence of a source of excitation, the wave structure of the basic stationary solutions will be perturbed and become weakly space-time variable. These wave-packet (or local stationary wave) solutions may arise in certain regions of space-time or at transition regions characterized by a rapid change of the amplitude and/or its derivatives. *Such regions do not arise in linear, homogeneous, stationary systems.* On the other hand, a rapid change of the wavelengths and/or frequencies do occur in the presence of spatial or temporal discontinuities (illustrated in Chapters 2 & 3). Thus, inhomogeneities or discontinuities in linear systems create effects somewhat similar to amplitude discontinuities in nonlinear systems.

6.2 "Mode" types in nonlinear systems (Riemann invariants)

In actual physical problems, if stationary solutions exist, fields are composed of many "mode" types; each type is characterized by its own speed and amplitude structure. They may or may not be independent of one another and indeed are coupled in "collision" regions wherein changes in

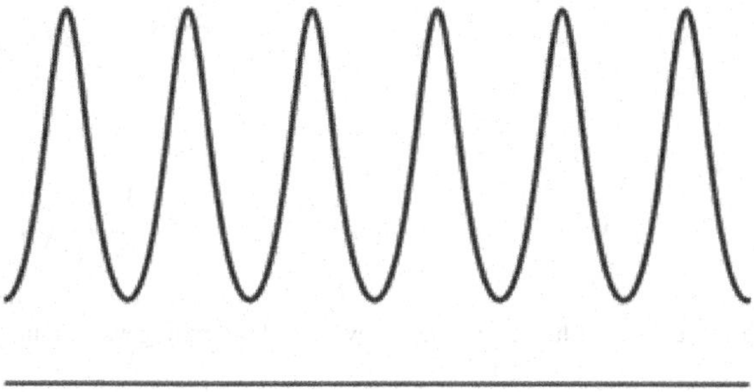

(a) Periodic, non-sinusoidal (Jacobi elliptic function-many sinusoidal harmonics).

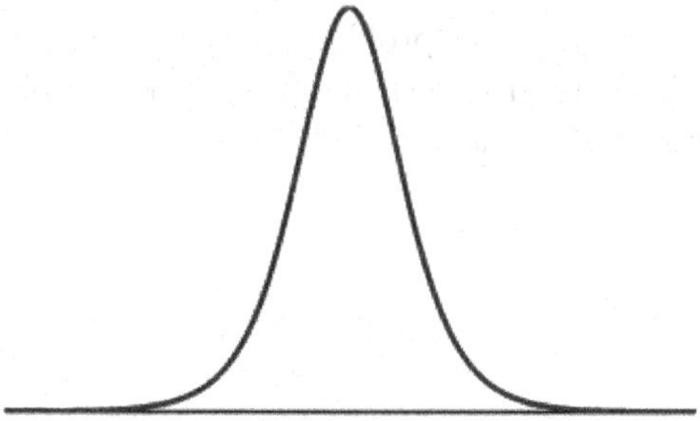

(b) Single soliton.

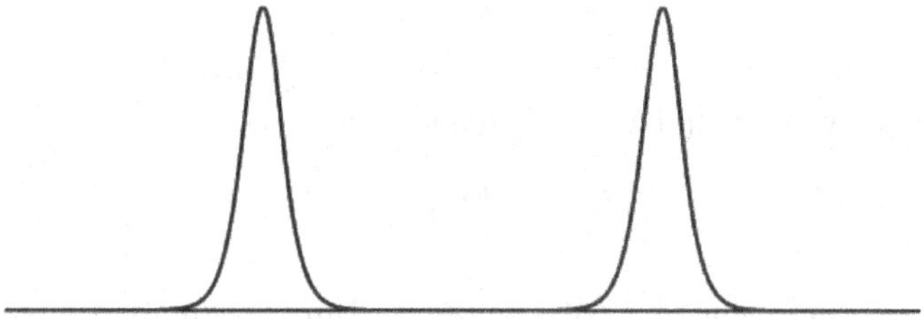

(c) Two solitons.

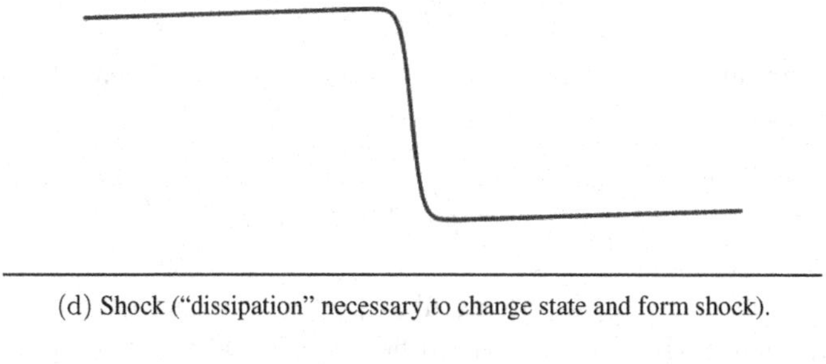

(d) Shock ("dissipation" necessary to change state and form shock).

Figure 6.1. Periodic and aperiodic wave solutions for nonlinear systems.

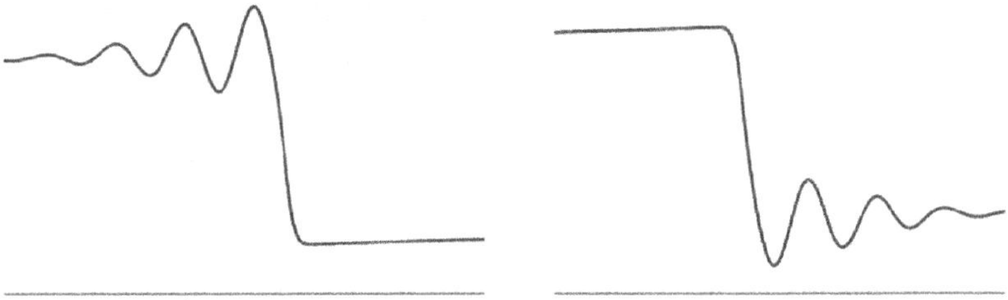

(e) Oscillatory (dispersive) shocks with leading/trailing wave train.

Figure 6.1. (*Continued*)

parameters occur. The general problem is very complicated, and hence, one first tries to understand generic single "mode" problems described by typical field equations of the form:

1. Nonlinear and non-dispersive

$$\frac{\partial \phi}{\partial t} + \phi \frac{\partial \phi}{\partial x} = 0$$

2. Nonlinear and dispersive (Korteweg-de Vrics equation)

$$\frac{\partial \phi}{\partial t} + \phi \frac{\partial \phi}{\partial x} + \alpha \frac{\partial^3 \phi}{\partial x^3} = 0$$

3. Nonlinear with damping but no dispersion

$$\frac{\partial \phi}{\partial t} + \phi \frac{\partial \phi}{\partial x} + \alpha |\phi|^2 = 0$$

4. Nonlinear and dispersive with damping (Burgers equation)

$$\frac{\partial \phi}{\partial t} + \phi \frac{\partial \phi}{\partial x} + \alpha \frac{\partial^2 \phi}{\partial x^2} = 0$$

5. Nonlinear and dispersive without damping (Nonlinear Schrödinger equation)

$$i \left(\frac{\partial \phi}{\partial t} + \mathrm{v} \frac{\partial \phi}{\partial x} \right) + \alpha \frac{\partial^2 \phi}{\partial x^2} + \beta |\phi|^2 \phi = 0$$

Nonlinearities usually occur at high power levels so that the material media supporting the waves are modified considerably by the propagating

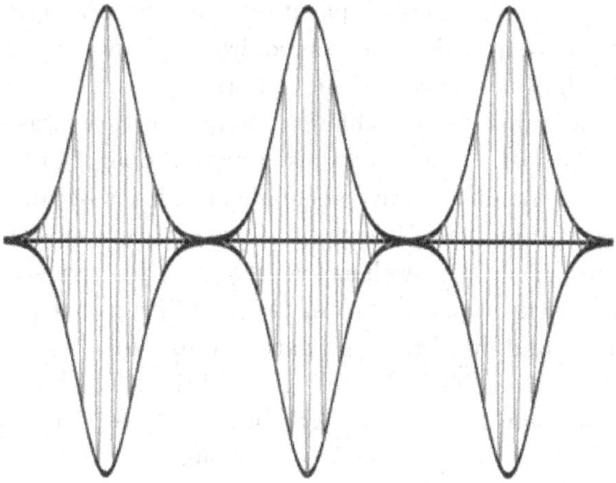

(a) Periodic modulation.

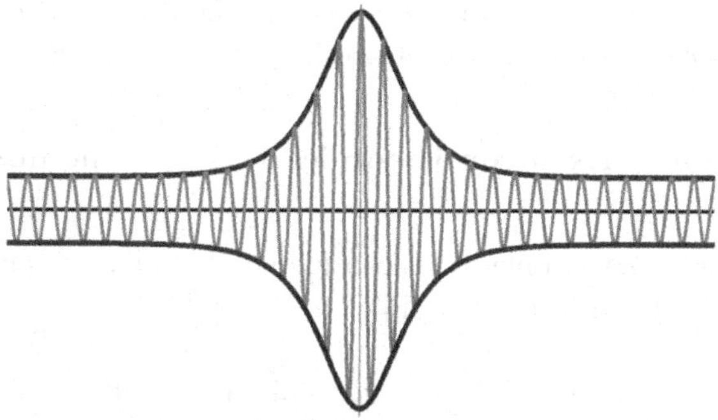

(b) Soliton modulation.

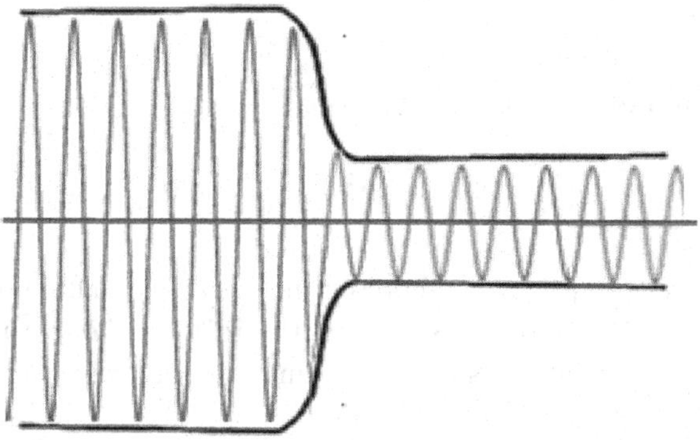

(c) Shock modulation.

Figure 6.2. High frequency nonlinear wave phenomena.

waves; consequently, the wave propagation and its characteristics undergo self-modification. Hence, the analysis requires a composite, self-consistent, descriptions both of the wave and the material systems.

Although the propagation of different nonlinear waves was studied originally in different media, many features of wave propagation in plasma can mimic most of the nonlinear wave phenomena. This is because plasma is a nonlinear dielectric medium; it supports high frequency EM and electron plasma (Langmuir) waves as well as low frequency ion acoustic wave. High frequency wave packets contain carriers; the nonlinear wave phenomena are revealed in the evolution of the modulation amplitudes (envelopes), which are governed by the nonlinear Schrödinger equation. The baseband low frequency waves are governed by the Korteweg de Vries (KdV) equation, in the study of nonlinear dispersive wave phenomena in collisionless plasma, and by the Burgers equation, in the study of nonlinear dissipative wave phenomena in collision plasma. These equations are first derived in the following. A selective overview of the aspects on and the analytical techniques used in nonlinear wave phenomena will be presented in Chapter 7.

6.3 Equations for self-consistent description of nonlinear waves in plasma

A uniform unmagnetized plasma is considered. Its fluid dynamic is defined, with the aid of the ideal gas law, $\mathrm{P_a = n_a T_a}$, by the

(1) continuity equations of the electron fluid and ion fluid

$$\frac{\partial \mathrm{n_a}}{\partial \mathrm{t}} + \boldsymbol{\nabla} \cdot \mathrm{n_a} \mathbf{v_a} = 0 \tag{6.1}$$

and

(2) electron and ion momentum equations

$$\mathrm{n_a m_a} \left(\frac{\partial \mathbf{v_a}}{\partial \mathrm{t}} + \mathbf{v_a} \cdot \boldsymbol{\nabla} \mathbf{v_a} \right) = -\boldsymbol{\nabla} \mathrm{P_a} + \mathrm{q_a n_a} \mathbf{E} \tag{6.2}$$

where the subscript a = e, i representing electron/ion fluid, respectively; $\boldsymbol{\nabla}\mathrm{P_a} = \gamma \mathrm{T_a} \boldsymbol{\nabla} \mathrm{n_a}$ and $\gamma = \mathrm{C_p/C_v}$ is the ratio of the specific heats $\mathrm{C_p}$ and $\mathrm{C_v}$, at constant pressure and volume, respectively. In the adiabatic compression, $\mathrm{P}/n^\gamma = $ Const., where $\gamma = (\mathrm{D}+2)/\mathrm{D}$, and D is the number of dimensions of the compression, i.e., $\gamma = 3, 2$, and $5/3$ for 1-, 2-, and 3-D compression. In the isothermal case, $\gamma = 1$.

The electric field in the momentum equations (6.2), are governed by Maxwell's equations

$$\nabla \times \mathbf{E} = -\frac{\partial \mathbf{B}}{\partial t} \tag{6.3a}$$

$$\nabla \times \mathbf{B} = \mu_0 \mathbf{J} + \frac{1}{c^2}\frac{\partial \mathbf{E}}{\partial t} \tag{6.3b}$$

$$\nabla \cdot \mathbf{E} = \frac{\rho}{\epsilon_0} \tag{6.3c}$$

$$\nabla \cdot \mathbf{B} = 0 \tag{6.3d}$$

where $\mathbf{J} = e(n_i \mathbf{v}_i - n_e \mathbf{v}_e)$ and $\rho = e(n_i - n_e)$ are the induced current density and charge density by the electric field $\mathbf{E}$ in plasma; singly charged ions are assumed; ϵ_0 is the free space permittivity. These two physical quantities ρ and $\mathbf{J}$ are related through the continuity (conservation of charge) equation

$$\frac{\partial \rho}{\partial t} + \nabla \cdot \mathbf{J} = 0 \tag{6.4}$$

The first two curl equations (6.3a) and (6.3b) are associated with the Faraday's law and Ampere's Law and the next two divergence equations (6.3c) and (6.3d) are the Gauss's law for the electric charges and magnetic charges.

6.4 Formulation of nonlinear wave equations

6.4.1 *Nonlinear Schrödinger equation for electromagnetic wave*

Equations (6.3a) and (6.3b) are combined to become

$$\frac{\partial^2}{\partial t^2}\mathbf{E} - c^2 \nabla^2 \mathbf{E} = -\frac{1}{\epsilon_0}\frac{\partial}{\partial t}\mathbf{J}_e \tag{6.5}$$

where $\mathbf{J}_e = \mathbf{J}_{eL} + \mathbf{J}_{eN} = -e(n_0 + \delta n)\mathbf{v}_{eL}$ is the induced electron current density by the wave field, $\mathbf{v}_{eL}$ is the electron linear velocity response to the wave field, and δn is the electron density perturbation induced by the radiation pressure of the wave modulation. With the aid of (6.2),

$$\frac{\partial}{\partial t}\mathbf{J}_{eL} = -en_0 \frac{\partial}{\partial t}\mathbf{v}_{eL} = \epsilon_0 \omega_p^2 \mathbf{E} \tag{6.5a}$$

and

$$\delta n = -\frac{n_0}{2}\frac{\langle|\mathbf{v}_{eL}|^2\rangle}{v_{te}^2}$$

where the electron thermal speed $v_{te} = \sqrt{T_e/m_e}$.

Consider a forward propagating wave modulation, which has a dominant carrier at (ω_1, k_1), a group velocity $v_g = k_1c^2/\omega_1$, and a modulation amplitude $\psi(\mathbf{r}, t)$, where

$$\omega_1 = \sqrt{\omega_p^2 + k_1^2c^2}$$

given by the linear dispersion relation (3.1a). Substitute $E(\mathbf{r}, t) = \psi(\mathbf{r}, t)e^{i(\mathbf{k}_1\cdot\mathbf{r}-\omega_1 t)} + \text{c.c.}$ and $v_{eL} \sim -i\frac{e}{m_e\omega_1}\psi e^{i(\mathbf{k}_1\cdot\mathbf{r}-\omega_1 t)} + \text{c.c.}$ in (6.2), yields

$$\delta n \cong -n_0\left(\frac{e}{m_e\omega_1}\right)^2\frac{|\psi|^2}{v_{te}^2} \tag{6.5b}$$

Substitute (6.5a) and (6.5b) into (6.5) and consider forward propagation in z (take forward wave approximation to neglect $\partial^2/\partial t^2$ term), it simplifies (6.5) to be

$$-i\left(\frac{\partial}{\partial t} + v_g\frac{\partial}{\partial z}\right)\psi(z, t) - \frac{1}{2}\frac{v_g}{k_1}\frac{\partial^2}{\partial z^2}\psi(z, t) - \frac{\omega_p^2}{2\omega_1 v_{te}^2}\left(\frac{e}{m_e\omega_1}\right)^2|\psi|^2\psi(z, t) = 0 \tag{6.5c}$$

In a moving frame at velocity $V = v_g$, and introduce dimensionless coordinate and time: $\xi = k_1(z - v_g t) = k_1 z - \tau$ and $\tau = k_1 v_g t$, and normalized wave function $\varphi(\xi, \tau) = \frac{\psi(z,t)}{\psi_0}$, where ψ_0 is the wave amplitude. With the aid of

$$\frac{\partial}{\partial t} = k_1 v_g\left(\frac{\partial}{\partial \tau} - \frac{\partial}{\partial \xi}\right), \quad \frac{\partial}{\partial z} = k_1\frac{\partial}{\partial \xi}, \quad \text{and} \quad \frac{\partial^2}{\partial z^2} = k_1^2\frac{\partial^2}{\partial \xi^2},$$

(6.5c) is converted to a one-dimensional nonlinear Schrödinger equation

$$-{}^1\!/\!{}_2\frac{\partial^2}{\partial \xi^2}\varphi - \alpha|\varphi|^2\varphi = i\frac{\partial}{\partial \tau}\varphi \tag{6.6}$$

where

$$\alpha = \frac{1}{2}\left(\frac{e}{m_e\omega_1 c}\right)^2\left(\frac{\omega_p}{k_1 v_{te}}\right)^2|\psi_0|^2.$$

On the LHS of (6.6), the first term induces wave dispersion and the second term steepens the wave, in the propagation.

In the linear case, i.e., $\alpha = 0$, (6.6) reduces to the case (c) of (1.16).

6.4.2 *Nonlinear Schrödinger equation for electron plasma (Langmuir) wave*

Electron plasma wave is an electrostatic wave, i.e., $\mathbf{E}_\ell = -\nabla\phi$, where ϕ is a scalar potential; thus only (6.3c) in the Maxwell's equations is involved in the formulation and becomes

$$\nabla \cdot \mathbf{E}_\ell = -\frac{\mathrm{e}(n_e - \hat{n})}{\epsilon_0} = -\frac{\mathrm{e}\delta n_{e\ell}}{\epsilon_0} \tag{6.7a}$$

where $\mathrm{n_e} = \mathrm{n_0} + \delta\mathrm{n}_{e\ell} + \mathrm{n_s}$; $\hat{\mathrm{n}} = \mathrm{n_0} + \mathrm{n_s}$; $\mathrm{n_0}$, $\delta\mathrm{n}_{e\ell}$, and $\mathrm{n_s}$ are the unperturbed plasma density and electron density perturbations associated with Langmuir waves and low frequency oscillations, respectively. Apply the operation $\partial/\partial\mathrm{t}$ to (6.1), i.e., $(\partial/\partial\mathrm{t})$ (6.1), and with the aid of (6.2) in which the convective term $(\mathbf{v}_\mathrm{e} \cdot \nabla\mathbf{v}_\mathrm{e})$ is neglected, it yields

$$\frac{\partial^2}{\partial \mathrm{t}^2}\delta n_{e\ell} - 3\mathrm{v}_{\mathrm{te}}^2\nabla^2\delta n_{e\ell} = (\mathrm{e}/\mathrm{m_e})[n_0\nabla \cdot \mathbf{E}_\ell + \nabla \cdot (n_s\mathbf{E}_\ell)] \tag{6.7b}$$

With the aid of (6.7a), a nonlinear mode equation governing the evolution of the Langmuir wave field $\mathbf{E}_\ell$ is derived to be

$$\left[\frac{\partial^2}{\partial \mathrm{t}^2} + \omega_\mathrm{p}^2 - 3\mathrm{v}_{\mathrm{te}}^2\nabla^2\right]\mathbf{E}_\ell = -\omega_\mathrm{p}^2\frac{n_s}{n_0}\mathbf{E}_\ell \tag{6.7c}$$

The nonlinear nature of the equation is shown implicitly by the RHS term of (6.7c), in which the wave induced background density perturbation, $\mathrm{n_s}$, modifies the dispersion property of the Langmuir wave self-consistently. The governing equation of this density perturbation is derived in the following. We combine the momentum equations of electron and ion fluids by adding them together. The electric field terms in the two equations cancel each other; and the electron inertial term $\mathrm{m_e}\partial\mathbf{v}_\mathrm{e}/\partial\mathrm{t}$ and the ion convective term $\mathrm{m_i}\mathbf{v}_\mathrm{i} \cdot \nabla\mathbf{v}_\mathrm{i}$, which are small comparing to their respective counterpart, are neglected. The combined equation is obtained to be

$$\mathrm{m_i}\frac{\partial}{\partial \mathrm{t}}\mathbf{v}_\mathrm{i} + \mathrm{m_e}\ \mathbf{v}_\mathrm{e} \cdot \nabla\ \mathbf{v}_\mathrm{e} = -\frac{1}{\mathrm{n_0}}\nabla(\mathrm{P_e} + \mathrm{P_i}) \tag{6.7d}$$

Apply the operation $(\nabla\cdot)$ to (6.7d), i.e., $(\nabla\cdot)$ (6.7d), and with the aid of the quasi-neutrality $\mathrm{n_{se}} = \mathrm{n_{si}} = \mathrm{n_s}$, and the relations $\mathbf{v}_\mathrm{e} \cdot \nabla\ \mathbf{v}_\mathrm{e} = \nabla(\mathrm{v}_\mathrm{e}^2/2)$

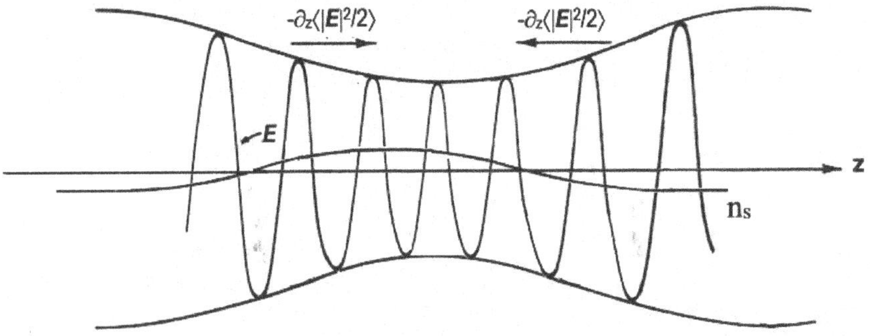

Figure 6.3. Density perturbation n_s set up by ponderomotive forces.

and $\partial n_s/\partial t + \nabla \cdot (n_0 \mathbf{v}_i) = 0$ deduced from the continuity equation, (6.7d) becomes

$$\left[\frac{\partial^2}{\partial t^2} - C_s^2 \nabla^2\right]\left(\frac{n_s}{n_0}\right) = \frac{m_e}{m_i}\nabla^2 \left\langle \frac{v_e^2}{2} \right\rangle \tag{6.7e}$$

where the implicit nonlinear term on the RHS of (6.7e) is attributed to a ponderomotive force acting on the electron plasma.

This force results from the non-uniform electron quiver motion in the Langmuir wave fields; as the electron plasma is pushed by this force, the induced electric field pulls ion plasma to move together to setup background density perturbation n_s as illustrated in Fig. 6.3.

This density perturbation varies slowly in time so that the time derivative term on the LHS of (6.7e) can be neglected. Thus, (6.7e) leads to

$$\frac{n_s}{n_0} \sim -\frac{m_e}{m_i}\left\langle \frac{v_{e\ell}^2}{2C_s^2} \right\rangle \tag{6.7f}$$

where $C_s = [(T_e + 3T_i)/m_i]^{1/2}$ is the ion acoustic speed. Again, consider a modulated wave forward propagating in z, which has a dominant carrier at (ω_1, k_1), a group velocity $v_{g\ell} = 3k_1 v_{te}^2/\omega_1$, and a modulation amplitude $\psi_\ell(z,t)$, where

$$\omega_1 = \sqrt{\omega_p^2 + 3k_1^2 v_{te}^2}$$

given by the linear dispersion relation of the Langmuir wave. Substitute $E_\ell(z,t) = \psi_\ell(z,t)e^{i(k_1 z - \omega_1 t)} + \text{c.c.}$ in (6.2), obtains

$$v_{e\ell} \sim -i\frac{e}{m_e\omega_1}\psi_\ell e^{i(k_1 z - \omega_1 t)} + \text{c.c.},$$

which is then substituted into (6.7f) to give

$$\frac{n_s}{n_0} \sim -\frac{m_e}{m_i}\left(\frac{e}{m_e\omega_1}\right)^2 \frac{|\psi_\ell|^2}{C_s^2} \tag{6.7g}$$

Apply the same procedure employed in Sec. **A** by substituting $E_\ell(z,t) = \psi_\ell(z,t)e^{i(k_1 z-\omega_1 t)}$+c.c. into (6.7c) and introducing dimensionless coordinate and time: $\xi_1 = k_1(z - v_{g\ell}t) = k_1 z - \tau_1$ and $\tau_1 = k_1 v_{g\ell}t$, in a moving frame at velocity $V = v_{g\ell}$, and normalizing the wave function:

$$\psi_\ell(\xi_1, \tau_1) = \frac{\psi_\ell(z,t)}{\psi_{\ell 0}},$$

(6.7c) is converted to a similar (to (6.6)) one dimensional nonlinear Schrödinger equation

$$-{}^1\!/_2\frac{\partial^2}{\partial\xi_1^2}\varphi_\ell - \alpha_1|\varphi_\ell|^2\varphi_\ell = i\frac{\partial}{\partial\tau_1}\varphi_\ell \tag{6.8}$$

where

$$\alpha_1 = \frac{1}{6}\left(\frac{e}{m_e\omega_1}\right)^2\left(\frac{\omega_{pi}}{k_1 v_{teC_s}}\right)^2|\psi_{\ell 0}|^2.$$

6.4.3 *Korteweg-de Vries (KdV) equation for ion acoustic wave*

Equations (6.1) and (6.2) are based to formulate the propagation of ion waves along z axis. Without the assumption of quasi-neutrality, (6.3c) is included to relate the wave field to the density oscillations. Add the electron and ion momentum equations and neglect the electron inertial terms, a one fluid equation is obtained to be

$$\left[\frac{\partial}{\partial t}V_{si} + \frac{\partial}{\partial z}\left(\frac{V_{si}^2}{2}\right)\right] = -C_s^2\frac{\partial}{\partial z}\left(\frac{\partial n_{si}}{n_0}\right) - \left(\frac{T_e}{m_i}\right)\frac{\partial}{\partial z}\left(\frac{\delta n_{se} - \delta n_{si}}{n_0}\right) \tag{6.9a}$$

where δn_{se} and δn_{si} are the electron and ion density perturbation and V_{si} is the velocity perturbation of the ion fluid in the presence of an ion wave. From the electron momentum equation (with the inertial terms neglected), it gives

$$E_s \sim -\frac{T_e}{n_0 e}\frac{\partial}{\partial z}\delta n_{se} \sim -\frac{T_e}{n_0 e}\frac{\partial}{\partial z}\delta n_{si},$$

where E_s is the self-consistent field induced in the ion wave to keep ambipolar motion (i.e., $V_{se} \sim V_{si}$). With the aid of the Gauss's law (6.3c):

$$(\delta n_{se} - \delta n_{si}) = -\frac{\epsilon_0}{e}\frac{\partial}{\partial z}E_s \sim \frac{v_s}{\omega_{pi}}\frac{\partial^2}{\partial z^2}\delta n_{si},$$

(6.9a) becomes

$$\left[\frac{\partial}{\partial t}V_{si} + \frac{\partial}{\partial z}\left(\frac{V_{si}^2}{2}\right)\right] = -C_s^2\frac{\partial}{\partial z}\left(\frac{\delta n_{si}}{n_0}\right) - \left(\frac{v_s^4}{\omega_{pi}^2}\right)\frac{\partial^3}{\partial z^3}\left(\frac{\delta n_{si}}{n_0}\right) \tag{6.9b}$$

where $v_s^2 = \frac{T_e}{m_i}$ and $\frac{\omega_{pi}^2}{v_s^2} = k_{De}^2$. Take the partial time derivative operation $\frac{\partial}{\partial t}$ on both sides of (6.9b) and with the aid of the ion continuity equation $\frac{\partial}{\partial t}\frac{\delta n_{si}}{n_0} = -\frac{\partial}{\partial z}V_{si}$, (6.9b) becomes

$$\frac{\partial}{\partial t}\left[\frac{\partial}{\partial t}V_{si} + \frac{\partial}{\partial z}\left(\frac{V_{si}^2}{2}\right)\right] = C_s^2\frac{\partial^2}{\partial z^2}V_{si} + \left(\frac{v_s^4}{\omega_{pi}^2}\right)\frac{\partial^4}{\partial z^4}V_{si} \tag{6.9c}$$

In the linear regime, the second (nonlinear) term in the LHS bracket and the second (dispersion) term on the RHS of (6.9c) are neglected; then, (6.9c) reduces to $\left(\frac{\partial^2}{\partial t^2} - C_s^2\frac{\partial^2}{\partial z^2}\right)V_{si} = 0$, which defines the linear dispersion relation $\omega_s = kC_s$ for the ion acoustic wave.

Equation (6.9c) is transformed to a moving frame at the ion acoustic velocity C_s by setting

$$t_1 = t \quad \text{and} \quad z_1 = z - C_s t$$

and letting $V_{si}(z,t) = A_s(z_1, t_1)$; thus

$$\frac{\partial}{\partial t} \to \frac{\partial}{\partial t_1} - C_s\frac{\partial}{\partial z_1} \quad \text{and} \quad \frac{\partial}{\partial z} \to \frac{\partial}{\partial z_1},$$

and (6.9c) becomes

$$\frac{\partial}{\partial t_1}\left[\frac{\partial}{\partial t_1}A_s + \frac{\partial}{\partial z_1}\left(\frac{A_s^2}{2}\right)\right] - 2C_s\frac{\partial}{\partial z_1}\left[\frac{\partial}{\partial t_1}A_s + \frac{\partial}{\partial z_1}\left(\frac{A_s^2}{4}\right)\right]$$
$$= \left(\frac{v_s^4}{\omega_{pi}^2}\right)\frac{\partial^4}{\partial z_1^4}A_s \tag{6.9d}$$

In the moving frame, the frequencies of the linear ion acoustic waves are downshifted to zero, thus, $\left|\frac{\partial}{\partial t_1}\right| \ll \left|C_s\frac{\partial}{\partial z_1}\right|$ and the first term on the LHS of (6.9d) is neglected. Introduce dimensionless variables and function:

$$\tau = \omega_{\rm pi} t_1, \eta = \beta z_1, \quad \text{and} \quad \phi(\eta, \tau) = \alpha_2 A_s(z_1, t_1),$$

where

$$\beta = \left(\frac{2C_s\omega_{\rm pi}^3}{v_s^4}\right)^{1/3} = \left(\frac{2C_s}{v_s}\right)^{1/3} k_{\rm De} \quad \text{and}$$

$$\alpha_2 = \frac{\beta}{12\omega_{\rm pi}} = \frac{1}{12v_s}\left(\frac{2C_s}{v_s}\right)^{1/3},$$

(6.9d) is converted to a standard KdV equation

$$\frac{\partial}{\partial\tau}\phi + \frac{\partial^3}{\partial\eta^3}\phi + 6\phi\frac{\partial}{\partial\eta}\phi = 0 \tag{6.10}$$

The first term characterizes the time evolution rate of the wave propagating under the influence of dispersion effect (second term) and the nonlinear steepening effect (third term) of the medium. In the linear regime, i.e., the third term on the LHS of (6.10) is neglected, it reduces to the case (d) of (1.16).

6.4.4 *Burgers equation for dissipated ion acoustic wave*

In collision plasma, collision terms are added to (6.2); one in the form of $-n_a m_a\, \nu_{ab}(v_a - v_b)$ is ascribed to collisions between electrons and ions, and the other one in the form of $-n_a m_a\, \nu_a v_a$ ascribed to collisions each other among the same species and with the background neutral particles. Because of momentum conservation, it imposes that $n_a m_a\, \nu_{ab} = n_b m_b\, \nu_{ba}$; thus in the one fluid equation, the terms associated with electron-ion collisions cancel each other. Quasi-neutrality is assumed, i.e., $\delta n_{\rm se} = \delta n_{\rm si}$, and the remaining collision terms representing the net collision damping effect on the ion wave; in fact, such collisions lead the diffusion of the perturbations, which in turn, causes ion wave to damp. Thus, the remaining collision terms in the one fluid equation are modelled by a single diffusion term in the form of $D\frac{\partial^2}{\partial z^2}V_{\rm si}$, where $D = \frac{\nu_i v_{\rm ti}^2}{\omega_{\rm pi}^2}$, and (6.9a) is modified to be

$$\left[\frac{\partial}{\partial t}V_{\rm si} + \frac{\partial}{\partial z}\left(\frac{V_{\rm si}^2}{2}\right)\right] = -C_s^2\frac{\partial}{\partial z}\left(\frac{\delta n_{\rm si}}{n_0}\right) + D\frac{\partial^2}{\partial z^2}V_{\rm si} \tag{6.11a}$$

Next, take the partial time derivative operation $\frac{\partial}{\partial t}$ on both sides of (6.11a) and apply the ion continuity equation $\frac{\partial}{\partial t}\frac{\delta n_{si}}{n_0} = -\frac{\partial}{\partial z}V_{si}$, it becomes

$$\frac{\partial}{\partial t}\left[\frac{\partial}{\partial t}V_{si} + \frac{\partial}{\partial z}\left(\frac{V_{si}^2}{2}\right)\right] = C_s^2\frac{\partial^2}{\partial z^2}V_{si} + D\frac{\partial}{\partial t}\frac{\partial^2}{\partial z^2}V_{si} \tag{6.11b}$$

Again, similar to (6.9c), (6.11b) is transformed to a moving frame at the ion acoustic velocity C_s by setting

$$t_1 = t \quad \text{and} \quad z_1 = z - C_s t$$

and letting $V_{si}(z,t) = A_{s1}(z_1, t_1)$; thus

$$\frac{\partial}{\partial t} \to \frac{\partial}{\partial t_1} - C_s\frac{\partial}{\partial z_1} \quad \text{and} \quad \frac{\partial}{\partial z} \to \frac{\partial}{\partial z_1},$$

and (6.11b) becomes

$$\begin{aligned}\frac{\partial}{\partial t_1}&\left[\frac{\partial}{\partial t_1}A_{s1} + \frac{\partial}{\partial z_1}\left(\frac{A_{s1}^2}{2}\right) - D\frac{\partial^2}{\partial z_1^2}A_{s1}\right]\\ &- 2C_s\frac{\partial}{\partial z_1}\left[\frac{\partial}{\partial t_1}A_{s1} + \frac{\partial}{\partial z_1}\left(\frac{A_{s1}^2}{4}\right)\right]\\ &= -C_s\, D\frac{\partial^3}{\partial z_1^3}A_{s1}\end{aligned} \tag{6.11c}$$

In the moving frame, $\left|\frac{\partial}{\partial t_1}\right| \ll \left|C_s\frac{\partial}{\partial z_1}\right|$, the first term on the LHS of (6.11c) is neglected. Introduce the similar dimensionless variables as those introduced in the K-dV case and a normalized function:

$$\tau = \omega_{pi}t_1, \eta = \beta z_1, \quad \text{and} \quad \phi_1(\eta,\tau) = \alpha_3 A_{s1}(z_1, t_1),$$

where

$$\beta = \left(\frac{2C_s\omega_{pi}^3}{v_s^4}\right)^{\frac{1}{3}} = \left(\frac{2C_s}{v_s}\right)^{\frac{1}{3}} k_{De}$$

$$\text{and} \quad \alpha_3 = \frac{\beta}{2\omega_{pi}} = \frac{1}{2v_s}\left(\frac{2C_s}{v_s}\right)^{1/3} = 6\alpha_2,$$

(6.11c) is converted to a standard Burgers equation

$$\frac{\partial}{\partial \tau}\phi_1 + \phi_1\frac{\partial}{\partial \eta}\phi_1 = b\frac{\partial^2}{\partial \eta^2}\phi_1 \tag{6.12}$$

where $b = \frac{D}{2\omega_{pi}}\beta^2$. The linear term on the RHS of (6.12) introduces diffusion effect and the nonlinear convection term on the LHS (second term) steepens the wave. In the linear regime, i.e., the second term on the LHS of (6.12) is neglected, it reduces to the case (b) of (1.16).

Problems

P6.1. In a linear system with a prescribed potential distribution $V(\xi)$, the linear Schrödinger equation is

$$-{}^1\!/_2\frac{\partial^2}{\partial\xi^2}\varphi + V(\xi)\varphi = \mathrm{i}\frac{\partial}{\partial\tau}\varphi \tag{P6.1}$$

If $V(\xi)$ is a localized potential well, then there will be a finite number of bound states with discrete eigen-energies $E_n = -\alpha_n^2$; $n = 1, 2, \ldots N$, and a continuum of states with eigen-energies $E = k^2$. These eigen-energies determine the corresponding eigen functions asymptotically (i.e., functions in the region $\xi \to \pm\infty$ where $V \to 0$). Find the asymptotic $\varphi(\xi, \tau)$ of the bound and unbound states.

P6.2. Show that the nonlinear Schrödinger equation (6.8) indicates that $\int_{-\infty}^{\infty} |\varphi_\ell|^2 \, d\xi_1$ is conserved.

P6.3. Galilean invariance: if $\phi(\eta, \tau)$ is a solution of the K-dV equation (6.10), show that $\tilde{\phi}(\eta, \tau) = \phi(\eta - 6v\tau, \tau) + v$ is also a solution.

P6.4. The Burgers equation (6.12) has the initial condition: $\phi_1(\eta, 0) = a\eta + b$, find an explicit solution $\phi_1(\eta, \tau)$ to (6.12).

Chapter 7

Analytical Solutions of Nonlinear Wave Equations

In Chapter 6, nonlinear equations descriptive of wave propagation in plasma are formulated. Those include the nonlinear Schrödinger equation, Korteweg-deVries (K-dV) equation, and Burgers equation, which are generic equations also applicable for the description of the similar nonlinear wave phenomena in other media. In the following, analytical methods to solve the equations are introduced and analytical solutions are presented.

7.1 Nonlinear Schrödinger equation (NLSE)

Most of wave propagation in weakly nonlinear, dispersive, energy-preserving systems are canonically descriptive, in an appropriate limit, by the NLS equation. Specifically, the NLS equation describes the evolution of a wave packet in a weakly nonlinear and dispersive medium when dissipation can be neglected. It has been applied for the study of optical pulse propagation in nonlinear fibers, the phenomenon of self-focusing, and the conditions under which an electromagnetic beam can propagate without spreading in nonlinear media, etc.

7.1.1 *Characteristic features of solutions*

It is hard to find a general analytical solution of the NLSE (6.6), however, it will be a good idea to identify what are the rigid constraints (attributes) that the equation imposes on its solutions. Those involve conservation laws, symmetries, and invariances.

A. Conservation laws

The conservation laws are in the form of continuity equation: $\frac{\partial}{\partial\tau}N + \frac{\partial}{\partial\eta}J = 0$, where N is the density of a conserved quantity and J is the flux of this quantity. Based on (6.6), the first two conservation laws are illustrated in the following:

$$\frac{\partial}{\partial\tau}|\varphi|^2 + \frac{\partial}{\partial\xi}\left[\frac{i}{2}|\varphi|^2\frac{\partial}{\partial\xi}\left(ln\frac{\varphi^*}{\varphi}\right)\right] = 0 \tag{7.1a}$$

$$\frac{\partial}{\partial\tau}\left[\frac{i}{2}|\varphi|^2\frac{\partial}{\partial\xi}\left(ln\frac{\varphi^*}{\varphi}\right)\right] + \frac{\partial}{\partial\xi}\left(\left|\frac{\partial\varphi}{\partial\xi}\right|^2 - \frac{1}{2}\alpha|\varphi|^4 - \frac{1}{4}\frac{\partial^2}{\partial\xi^2}|\varphi|^2\right) = 0 \tag{7.1b}$$

The physical quantities are identified to be

$$N_1 = |\varphi|^2 \quad \text{and} \quad J_1 = \frac{i}{2}|\varphi|^2\frac{\partial}{\partial\xi}\left(ln\frac{\varphi^*}{\varphi}\right) = N_2$$

$$\begin{aligned} J_2 &= \left|\frac{\partial\varphi}{\partial\xi}\right|^2 - \frac{1}{2}\alpha|\varphi|^4 - \frac{1}{4}\frac{\partial^2}{\partial\xi^2}|\varphi|^2 \\ &= \frac{1}{2}\left(\left|\frac{\partial\varphi}{\partial\xi}\right|^2 - \alpha|\varphi|^4\right) - \frac{1}{4}\left(\varphi^*\frac{\partial^2}{\partial\xi^2}\varphi + \varphi\frac{\partial^2}{\partial\xi^2}\varphi^*\right) \end{aligned}$$

where $N_1, J_1, and\, J_2$ represent the mass density, momentum density, and pressure, respectively, of a unit mass particle system.

In addition, the Hamiltonian of the system is defined to be

$$\text{H} = 1/2\int_{-\infty}^{\infty}\left(\left|\frac{\partial\varphi}{\partial\xi}\right|^2 - \alpha|\varphi|^4\right)d\xi \tag{7.1c}$$

One can show that $\frac{dH}{d\tau} = 0$; thus, H is a constant of motion, and given by the initial value

$$\text{H} = 1/2\int_{-\infty}^{\infty}\left(\left|\frac{\partial\varphi_0}{\partial\xi}\right|^2 - \alpha|\varphi_0|^4\right)d\xi$$

B. Scaling symmetry

If $\varphi(x,t)$ is a solution of (6.6) with the initial condition: $\varphi(x,\,0) = \varphi_0(x)$, then

$$\varphi_\alpha(x,\,t) = \alpha^{-1}\varphi\left(\frac{x}{\alpha},\frac{t}{\alpha^2}\right) \tag{7.1d}$$

is also a solution of (6.6) with the initial condition: $\varphi_\alpha(x, 0) = \alpha^{-1}\varphi\left(\frac{x}{\alpha}, 0\right) = \alpha^{-1}\varphi_0\left(\frac{x}{\alpha}\right)$.

C. Galilean invariance

If $\varphi(x,t)$ is a solution of (6.6) with the initial condition: $\varphi(x,0) = \varphi_0(x)$, then

$$\varphi_k(x,\, t) = e^{ikx}e^{\frac{i|k|^2 t}{2}}\varphi(x - kt, t) \tag{7.1e}$$

is also a solution of (6.6) with the initial condition: $\varphi_k(x,0) = e^{ikx}\varphi(x,0) = e^{ikx}\varphi_0(x)$.

D. Virial theorem (Variance identity)

A variance $V(\tau)$ is defined to be

$$V(\tau) = \int_{-\infty}^{\infty} \xi^2|\varphi(\xi,\tau)|^2 d\xi \tag{7.1f}$$

Apply (6.6) and integration by parts, the variance identity is obtained to be

$$\frac{d^2}{d\tau^2}V(\tau) = 4H + \alpha\int |\varphi|^4 d\xi \tag{7.1g}$$

It is noted that $V(\tau) \geq 0$ for all τ. This equation will be generalized to the multi-dimensional cases in Chapter 8 to explore the conditions of wave collapse.

7.1.2 *Analyses*

The nonlinear Schrödinger equation (6.6) is analyzed by first converting it to an eigenvalue equation. It is done by introducing $\varphi(\xi,\tau) = \phi(\xi)e^{-iE\tau}$, where E is the eigenvalue of a state and $\phi(\xi)$ is a real function. An eigenvalue equation is obtained to be

$$\left(-{}^1\!/\!_2\frac{d^2}{d\xi^2} - \alpha\phi^2\right)\phi = \mathrm{E}\phi \tag{7.2}$$

Consider $(\phi,\ \xi)$ as the equivalent spatial coordinate and time of a system, Eq. (7.2) represents an equation of motion of a unit mass object moving in this one-dimensional space and being accelerated by a force of $-2(\mathrm{E}\phi+\alpha\phi^3)$. Multiply $\frac{d}{d\xi}\phi$ to both sides of (7.2), it leads to

$$\frac{d}{d\xi}\left[1/2\left(\frac{d\phi}{d\xi}\right)^2 + (\mathrm{E}\phi^2 + 1/2\alpha\phi^4)\right] = 0 \tag{7.3}$$

It is recognized that the quantity in the parenthesis on the LHS of (7.3) is invariant with ξ.

Hence, this unit mass object, moving in a potential field $\mathrm{V}(\phi) = \mathrm{E}\phi^2 + 1/2\,\alpha\phi^4$, has energy $H = 1/2\left(\frac{d\phi}{d\xi}\right)^2 + (\mathrm{E}\phi^2 + 1/2\,\alpha\phi^4)$, which is constant with the equivalent time ξ, where $\mathrm{T} = 1/2(\frac{d\phi}{d\xi})^2$ is the kinetic energy of this object. Two typical plots of the potential function in the cases of $\mathrm{E} > 0$ and $\mathrm{E} < 0$ are illustrated in Fig. 7.1(a). As shown, both plots represent potential wells.

In the case of $\mathrm{E} > 0$, the total energy H of the trapped object is larger than zero, i.e., $\mathrm{H} > 0$; the trapped object is bounced back and forth in the potential well to have an oscillatory trajectory $\phi_{\mathrm{p1}}(\xi)$, which is a symmetric alternate function illustrated in Fig. 7.1(b). In the case of $\mathrm{E} < 0$, the trapped object has $\mathrm{H} > 0$ or $\mathrm{E}\phi_{\mathrm{m}}^2 + 1/2\,\alpha\phi_{\mathrm{m}}^4 < \mathrm{H} < 0$. If $\mathrm{H} > 0$, the bounce motion of the object has an oscillatory trajectory $\phi_{\mathrm{p2}}(\xi)$, which is a symmetric alternate function illustrated in Fig. 7.1(b). If $\mathrm{E}\phi_{\mathrm{m}}^2 + 1/2\,\alpha\phi_{\mathrm{m}}^4 < \mathrm{H} < 0$, the oscillatory trajectory $\varphi_{\mathrm{p3}}(\xi)$ of the object is a non-alternate function as shown in Fig. 7.1(b). Moreover, there exists a non-oscillatory trajectory with $\mathrm{H} = 0$ in the case of $\mathrm{E} < 0$. Let the object start at $\phi = \phi_1$ $(= 0)$ and exam the motion of the object in the region between ϕ_1 and ϕ_2 in Fig. 7.1(a). Initially, it moves very slowly to the right. As it drops into the potential well, it moves quickly toward the potential minimum at φ_{m}. After passing the potential minimum, the object starts to climb up to the turning point at $\phi = \phi_2$, where the object is bounced back into the potential well. It quickly passes the potential minimum and then climbs up toward the starting point $\phi_1 = 0$. It takes a long time for the object to reach ϕ_1, where the object stays. This non-oscillatory trajectory $\phi_{\mathrm{s}}(\xi)$ is also shown in Fig. 7.1(b).

A. Periodic solutions

The potential distribution functions illustrated in Fig. 7.1(a) indicate that the solutions of (7.2) in the case of $\mathrm{H} \neq 0$ are periodic as illustrated

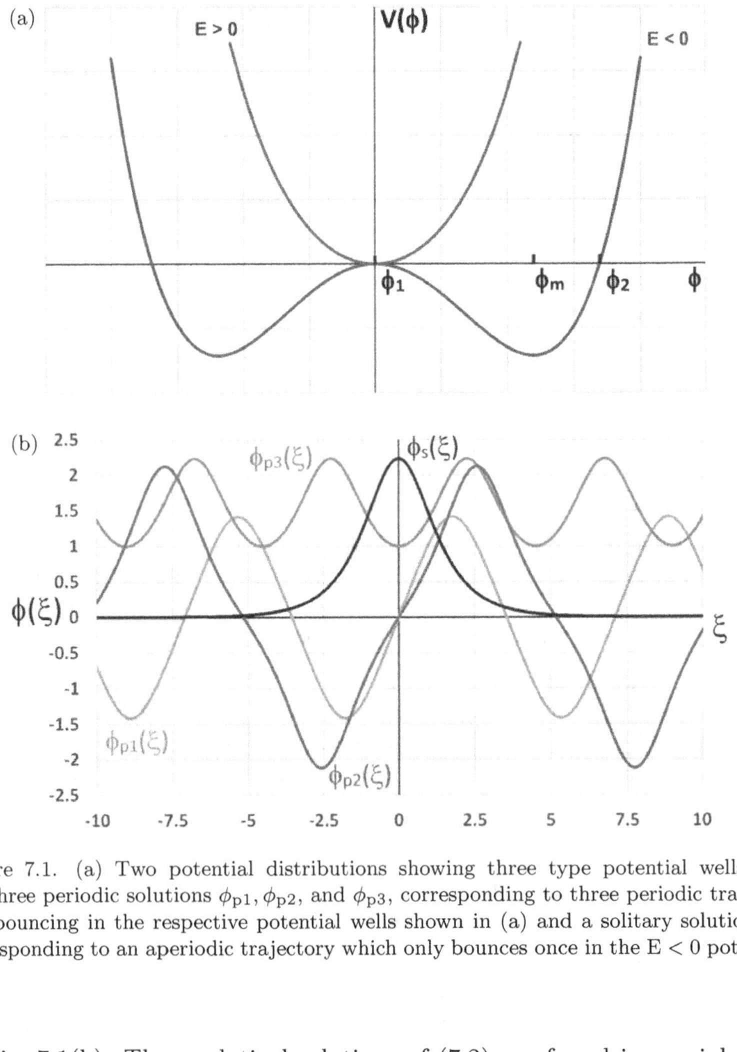

Figure 7.1. (a) Two potential distributions showing three type potential wells and (b) three periodic solutions ϕ_{p1}, ϕ_{p2}, and ϕ_{p3}, corresponding to three periodic trajectories bouncing in the respective potential wells shown in (a) and a solitary solution ϕ_s, corresponding to an aperiodic trajectory which only bounces once in the E < 0 potential well.

in Fig. 7.1(b). The analytical solutions of (7.2) are found in special cases as exemplified in the following.

1. For E > 0, i.e. H > 0. Set $\eta_1 = [2E/(1 - 2k_1^2)]^{1/2}\xi$, $\phi(\xi) = \phi_{10} y_1(\eta_1)$, and $k_1^2 = (1 - 2k_1^2)\frac{\alpha\phi_{10}^2}{2E} = \frac{1}{2}(1 - \frac{1}{\sqrt{1+2\alpha H/E^2}}) < \frac{1}{2}$, where $\phi_{10}^2 = \frac{E}{\alpha}(\sqrt{1 + 2\alpha H/E^2} - 1)$, (7.2) is normalized to be

$$y_1'' + (1 - 2k_1^2)y_1 + 2k_1^2 y_1^3 = 0 \tag{7.4a}$$

where $y_1'' = \frac{d^2}{d\eta_1^2}y_1$. The solution of (7.4a) is a Jacobi elliptic (cosine amplitude) function $cn\,(\eta_1, \mathrm{k}_1)$; thus,

$$\phi_{\mathrm{p}1}(\xi) = \phi_{10}\, cn\,(\eta_1, \mathrm{k}_1)$$

which is a symmetric alternate function as shown in Fig. 7.1(b).

2. For E < 0, H > 0. Again, set $\eta_2 = [2\mathrm{E}/(1 - 2\mathrm{k}_2^2)]^{1/2}\xi$, $\phi(\xi) = \phi_{20}\mathrm{y}_2(\eta_2)$, and $k_2^2 = (1 - 2k_2^2)\frac{\alpha\phi_{20}^2}{2E} = \frac{1}{2}(1 + \frac{1}{\sqrt{1+2\alpha H/E^2}}) > \frac{1}{2}$, where $\phi_{20}^2 = -\frac{E}{\alpha}(\sqrt{1 + 2\alpha H/E^2} + 1)$, (7.2) is normalized to be

$$y_2'' + (1 - 2k_2^2)y_2 + 2k_2^2 y_2^3 = 0 \tag{7.4b}$$

where $y_2'' = \frac{d^2}{d\eta_2^2}y_2$. Eq. (7.4b) has the same form as (7.4a), its solution is also a Jacobi elliptic function $cn\,(\eta_2, \mathrm{k}_2)$. Thus,

$$\phi_{\mathrm{p}2}(\xi) = \phi_{20}\, cn\,(\eta_2, \mathrm{k}_2);$$

likewise, it is a symmetric alternate function as shown in Fig. 7.1(b).

3. For E < 0 and H < 0. Set $\eta_3 = [-2\mathrm{E}/(2 - \mathrm{k}_3^2)]^{1/2}\xi$, $\phi(\xi) = \phi_{30}\mathrm{y}_3(\eta_3)$, and $-(2 - k_3^2)\frac{\alpha\phi_{30}^2}{E} = 2$, i.e., $k_3^2 = 2(1 + \frac{1}{\sqrt{1+2\alpha H/E^2}})^{-1}$, where $\phi_{30}^2 = -\frac{E}{\alpha}(\sqrt{1 + 2\alpha H/E^2} + 1)$, (7.2) is normalized to be

$$y_3'' - (2 - k_3^2)y_3 + 2y_3^3 = 0 \tag{7.4c}$$

where $y_3'' = \frac{d^2}{d\eta_3^2}y_3$. The solution of (7.4c) is a Jacobi elliptic (delta amplitude) function $dn\,(\eta_3, \mathrm{k}_3)$; thus,

$$\phi_{\mathrm{p}3}(\xi) = \phi_{30}\, dn\,(\eta_3, \mathrm{k}_3)$$

which is a non-alternate periodic function as shown in Fig. 7.1(b).

B. Solitary solution

A localized solution of (7.2) requires $\phi = 0 = \phi'$ as $|\xi| \to \infty$, thus, H = 0 in the case of E < 0 is considered. Set $\mathrm{x} = \sqrt{2|\mathrm{E}|}\xi$, $\phi(\xi) = \phi_{\mathrm{s}0}\mathrm{Y}_{\mathrm{s}}(\mathrm{x})$, and where $\phi_{\mathrm{s}0} = \sqrt{\frac{2|\mathrm{E}|}{\alpha}}$, (7.2) is normalized to be

$$\mathrm{Y}_{\mathrm{s}}'' - \mathrm{Y}_{\mathrm{s}} + 2\mathrm{Y}_{\mathrm{s}}^3 = 0 \tag{7.5}$$

where $Y_s'' = \frac{d^2}{dx^2} Y_s$. Compare (7.5) with (7.4b) and (7.4c), the solution of (7.5) is $cn\,(x, 1) = dn\,(x, 1) = \text{sech}\, x$. Thus

$$\phi_s(\xi) = \sqrt{\frac{2|E|}{\alpha}}\,\text{sech}\sqrt{2|E|}\xi \tag{7.5a}$$

This is a solitary solution that $\phi_s = 0 = \phi_s'$ as $|\xi| \to \infty$. Its width ($\propto \frac{1}{\sqrt{2|E|}}$) is inversely proportional to its amplitude. This solution is also plotted in Fig. 7.1(b).

The cubic nonlinearity of the medium mitigates wave dispersion in the propagation; when the nonlinear effect ($\propto$ square of the amplitude) and the dispersion effect (inversely proportional to the square of the width) reach a balance, a shape-preserved solitary wave is formed and trapped in the self-induced density well ($\propto -\phi_s^2$).

With the aid of (7.5a), the field function of a solitary wave packet is obtained to be

$$E_s(z,\, t) = \sqrt{\frac{2|E|}{\alpha_0}}\,\text{sech}[\sqrt{2|E|}k_1(z - v_g t)]\cos[k_1 z - (\omega_1 + \Delta\omega)t] \tag{7.6}$$

where $\alpha_0 = \frac{\alpha}{|\psi_0|^2}$ and $\Delta\omega = Ek_1 v_g < 0$. This envelope soliton (7.6) and the self-induced potential well ($\propto -\phi_s^2$) to trap this solitary wave are plotted in Fig. 7.2.

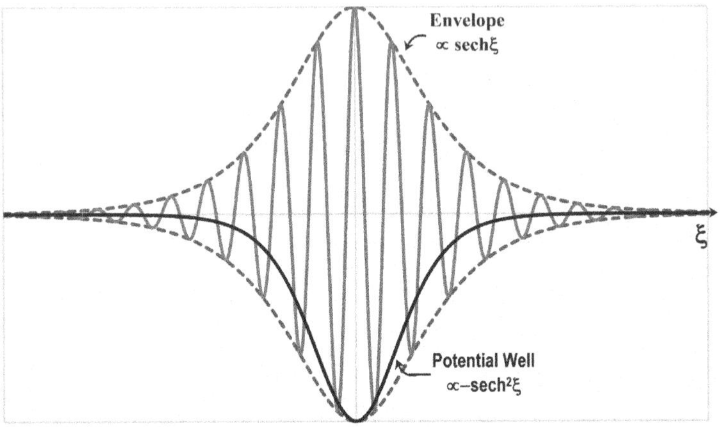

Figure 7.2. Envelope soliton solution of the nonlinear Schrödinger equation (6.6) and the self-induced potential well trapping the soliton.

With the aid of the relations:

$$\int_{-\infty}^{\infty} \mathrm{sech} z\, dz = \pi \quad \text{and} \quad \int_{-\infty}^{\infty} \mathrm{sech}^2 \mathrm{z}\, \mathrm{dz} = 2,$$

it is shown that the area $\int_{-\infty}^{\infty} \phi_s(\xi)d\xi = \frac{\pi}{\sqrt{\alpha}}$ of a soliton is independent of the amplitude; moreover, its energy $\int_{-\infty}^{\infty} \phi_s^2(\xi)\, d\xi = 2\frac{\sqrt{2|E|}}{\alpha}$ is linearly proportional to the amplitude. This is because the width of the soliton is inversely proportional to the amplitude. Both quantities are inversely proportional to the square root of the nonlinear coefficient α (i.e., $\propto \frac{1}{\sqrt{\alpha}}$).

The solution (7.6) indicates that plasma can support solitons, which are shape-preserved localized wave packets. However, it is realized that soliton is not a necessity of the plasma nonlinearity, the nonlinear plasma waves, in general, are periodic; soliton(s) may appear when the source wave function has a localized form.

7.2 Korteweg-de Vries (K-dV) equation

It models a variety of nonlinear phenomena, including ion acoustic waves in plasmas, and shallow water waves. In (6.10), the first term characterizes the time evolution of the wave propagating in one direction, the second term disperses the wave, and the third term steepens the wave.

7.2.1 *Conservation laws*

The K-dV equation (6.10) possess many conservation laws, which are in the form of continuity equation: $\frac{\partial}{\partial \tau}N + \frac{\partial}{\partial \eta}J = 0$, where N is the density of a conserved quantity and J is the flux of this quantity. The density N of the first two conservation laws are in the form of ϕ^{n}, for n = 1 and 2; in the higher order cases, N has more complicated form, e.g., $N = \phi^3 - \frac{1}{2}\phi_\eta^2$ in the third conservation law. In the following, the first three conservation laws are illustrated. Eq. (6.10) is rearranged to obtain the first conservation law

$$\frac{\partial}{\partial \tau}\phi + \frac{\partial}{\partial \eta}(3\phi^2 + \phi_{\eta\eta}) = 0 \tag{7.7a}$$

Next, multiplying (6.10) by ϕ, the second conservation law is obtained to be

$$\frac{\partial}{\partial \tau}\frac{\phi^2}{2} + \frac{\partial}{\partial \eta}\left[2\phi^3 + \phi\phi_{\eta\eta} - \frac{1}{2}\phi_\eta^2\right] = 0 \tag{7.7b}$$

The third conservation law is given as

$$\frac{\partial}{\partial\tau}\left(\phi^3-\frac{1}{2}\phi_\eta^2\right)+\frac{\partial}{\partial\eta}\left[\frac{9}{2}\phi^4+3\phi^2\phi_{\eta\eta}-6\phi\phi_\eta^2+\frac{1}{2}\phi_{\eta\eta}^2-\phi_\eta\phi_{\eta\eta\eta}\right]=0 \tag{7.7c}$$

Higher order conservation laws can be found in literatures. It is conjectured that the K-dV equation has infinite number of conservation laws.

7.2.2 *Potential and modified Korteweg-de Vries (p & mK-dV) equations*

There are several variants of the K-dV equation, which are derived via different function transformations:

1. Set $\phi(\eta,\tau)=\frac{\partial}{\partial\eta}\mathrm{U}_p(\eta,\tau)$ in (6.10), it leads to the "potential K-dV" equation

$$\frac{\partial}{\partial\tau}\mathrm{U}_p+\frac{\partial^3}{\partial\eta^3}\mathrm{U}_p+3\left(\frac{\partial}{\partial\eta}\mathrm{U}_p\right)^2=0 \tag{7.8}$$

2. Apply Miura transformation:

$$\phi=-\left(\mathrm{U}^2+\frac{\partial}{\partial\eta}\mathrm{U}\right) \tag{7.9}$$

to the K-dV equation (6.10), i.e., substitute (7.9) into (6.10), it shows that $\mathrm{U}(\eta,\tau)$ satisfies a modified K-dV (mK-dV) equation

$$\frac{\partial}{\partial\tau}\mathrm{U}+\frac{\partial^3}{\partial\eta^3}\mathrm{U}-6\mathrm{U}^2\frac{\partial}{\partial\eta}\mathrm{U}=0 \tag{7.10}$$

Both equations also possess many conservation laws.

7.2.3 *Propagating modes*

Consider traveling wave solution of the form $\phi(\eta,\tau)=\mathrm{f}(\xi)$, where $\xi=\eta-A_s\tau$, (6.10) becomes

$$\frac{\mathrm{d}^3}{\mathrm{d}\xi^3}\mathrm{f}-A_s\frac{\mathrm{d}}{\mathrm{d}\xi}\mathrm{f}+3\frac{\mathrm{d}}{\mathrm{d}\xi}f^2=0 \tag{7.11}$$

This equation is integrated to be

$$\frac{\mathrm{d}^2}{\mathrm{d}\xi^2}\mathrm{f}-A_s\mathrm{f}+3\mathrm{f}^2=\mathrm{C}_0 \tag{7.11a}$$

where the integration constant $C_0 = [f''(0) - A_s f(0) + 3f^2(0)]$. It is noted that C_0 can be cancelled by shifting f by a constant, i.e., $f = f_1 + C_1$ and $C_1 = \frac{A_s}{6}(1 - \sqrt{1 + 12C_0/A_s^2})$; on the other hand, C_1 has to be zero physically; hence, $C_0 = 0$. Multiply $\frac{d}{d\xi}f$ to (7.11a), yields an invariant equation

$$\frac{d}{d\xi}\left[\frac{1}{2}\left(\frac{d}{d\xi}f\right)^2 + \left(-\frac{A_s}{2}f^2 + f^3\right)\right] = 0 \tag{7.11b}$$

It indicates that the quantity in the parenthesis on the LHS of (7.11b) is invariant with ξ; that is

$$1/2 f'^2 + (-1/2\, A_s f^2 + f^3) = H \tag{7.11c}$$

where $f' = df/d\xi$ and the integration constant $H = 1/2[f'^2(0) - A_s f^2(0) + 2f^3(0)]$. Consider (f, ξ) as the coordinate and time of a one-dimensional space having a potential distribution $V(f) = (-1/2\, A_s f^2 + f^3)$, (7.11a) represents the equation of motion of an unit mass object moving in this potential field; this object has kinetic energy $KE = 1/2 f'^2$, potential energy $VE = (-1/2 A_s f^2 + f^3)$, and a constant total energy $TE = VE + KE = H$. This is equivalent to the physical interpretation of nonlinear oscillations presented in Sec. 5.4 (Fig. 5.1).

A typical plot of the potential field V(f), for $A_s = 2$, is illustrated in Fig. 7.3(a). It exhibits a potential well, which traps objects to force oscillations. The oscillatory trajectories convert to periodic solutions of (7.11). As an example, a periodic solution, for $H = -3/64$ and the initial conditions: $f(0) = 0.25$ and $f'(0) = 0$, is illustrated in Fig. 7.3(b). Moreover, there exists a non-oscillatory trajectory in the case of $H = 0$. It will be shown to be a solitary solution of (7.11).

A. Periodic Solution

Although (7.11) cannot be converted to a Jacobi elliptic equation, its periodic solution, under a special condition, can be obtained indirectly via the modified K-dV equation (7.10). For the traveling wave solution of the form $V(\eta, \tau) = g(\xi)$, where $\xi = \eta - A_s\tau$, (7.10) becomes

$$\frac{d^3}{d\xi^3}g - A_s\frac{d}{d\xi}g - 2\frac{d}{d\xi}g^3 = 0 \tag{7.12}$$

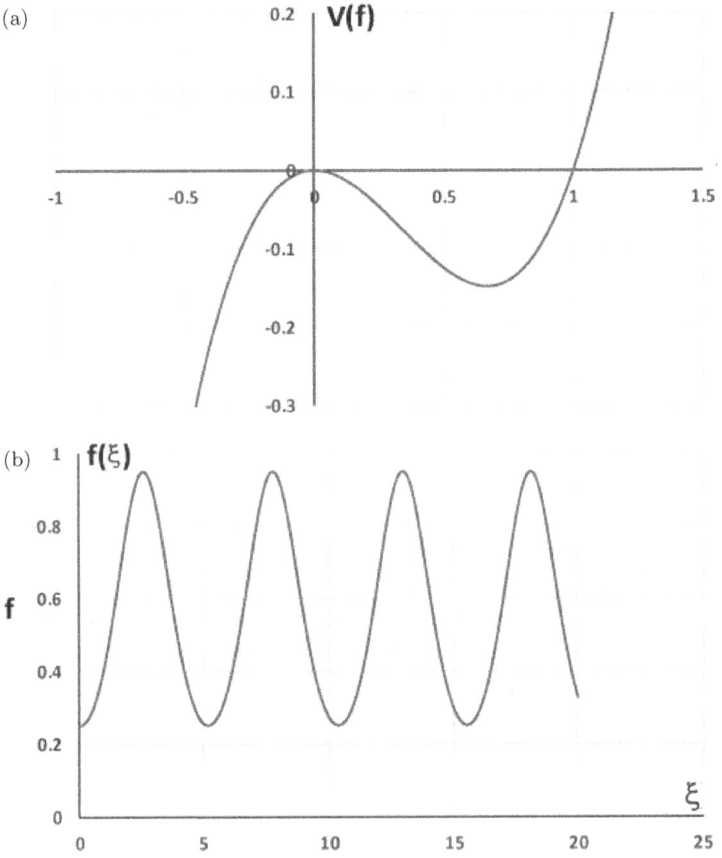

Figure 7.3. (a) A typical plot of the potential function V(f) and (b) a periodic solution of Eq. (7.11).

This equation is integrated to be

$$\frac{d^2}{d\xi^2}g - A_s g - 2g^3 = C \tag{7.12a}$$

Consider a special case that the integration constant $C = 0$ and set $g(\xi) = ky(\xi)$ and $A_s = -(1 + k^2)$, then (7.12a) is converted to a Jacobi elliptic equation

$$y'' + (1 + 2k^2)y - 2k^2y^3 = 0. \tag{7.12b}$$

Its solution is the Jacobi elliptic (sine amplitude) function $sn(\xi + \vartheta, k)$. Hence, the periodic solution of (7.11) in the case of C = 0 is obtained to be

$$\mathrm{f}(\xi;\mathrm{k}) = -[\mathrm{k}^2\,\mathrm{sn}^2(\xi+\vartheta,\mathrm{k}) + \mathrm{k}\,\mathrm{cn}(\xi+\vartheta,\mathrm{k})\,\mathrm{dn}(\xi+\vartheta,\mathrm{k})] \tag{7.12c}$$

where the constant parameters k and ϑ are determined by the initial conditions. Although this is not a realistic case because $\mathrm{f}(\xi;\mathrm{k}=1) = -1$, which is a trivial solution of (7.11), it is an example having an analytical solution. Since the potential functions of the K-dV and mK-dV equations are elliptic curves, elliptic function solutions are also expected in the general cases and can be obtained numerically.

B. Soliton trapped in self-induced potential well

Eq. (7.11c), with H = 0, leads to $\mathrm{f}' = -\mathrm{f}\sqrt{A_s - 2\mathrm{f}}$, which is then integrated to obtain a solitary solution of the form

$$\mathrm{f}(\xi) = \phi(\eta,\tau) = \phi(\eta - A_\mathrm{s}\tau) = {}^1\!/_2 A_\mathrm{s}\,\mathrm{sech}^2[{}^1\!/_2\sqrt{A_\mathrm{s}}(\eta - A_\mathrm{s}\tau)] \tag{7.13}$$

It is shown that the velocity (in the moving frame) of an ion acoustic soliton is proportional to its amplitude A_s and the width is inversely proportional to the square root of the amplitude. This is realized because, in (7.7a), the nonlinear effect is proportional to the amplitude and the dispersion effect is inversely proportional to the square of the width.

At the identified amplitude-width relationship, these two effects reach balance to form a shape-preserved soliton, which is trapped in the self-induced density well ($\propto -\phi_\mathrm{s}^2$). The area $\int_{-\infty}^{\infty}\phi(\eta,\tau)\mathrm{d}\eta = 2A_\mathrm{s}^{1/2}$ and energy $\int_{-\infty}^{\infty}\phi^2(\eta,\tau)d\eta = \frac{2}{3}A_\mathrm{s}^{3/2}$ of an ion acoustic soliton are amplitude dependent. These relations impose conditions on the source (initial) pulse which is likely to evolve nonlinearly to become a soliton as shown in Fig. 7.4, in which the self-induced ponderomotive force to balance the wave dispersion is also plotted.

An ion acoustic soliton can be seen as a traveling plasma density bump. Substitute (7.13) into the ion continuity equation of (6.1): $\frac{\partial}{\partial t}\frac{n_{si}}{n_0} + \frac{\partial}{\partial z}V_{si} = 0$, where $\mathrm{V_{si}} = \phi/\alpha_2$, and $\tau = \omega_\mathrm{pi}\mathrm{t}$ and $\eta = \beta(\mathrm{z} - \mathrm{C_s t})$, this traveling density bump has the form

$$\frac{\mathrm{n_{si}}}{\mathrm{n_0}} = \frac{12A_{s1}}{1+2A_{s1}}\,\mathrm{sech}^2\left[\left(\frac{\mathrm{C_s}}{\mathrm{v_s}}\right)\sqrt{A_{s1}}\mathrm{k_{De}}(\mathrm{z} - \mathrm{C_{s1}t})\right] \tag{7.13a}$$

where $A_{\mathrm{s}1} = (\mathrm{v_s}/2\mathrm{C_s})^{4/3}A_s$ and $C_{s1} = (1+2A_{s1})\mathrm{C}_s$.

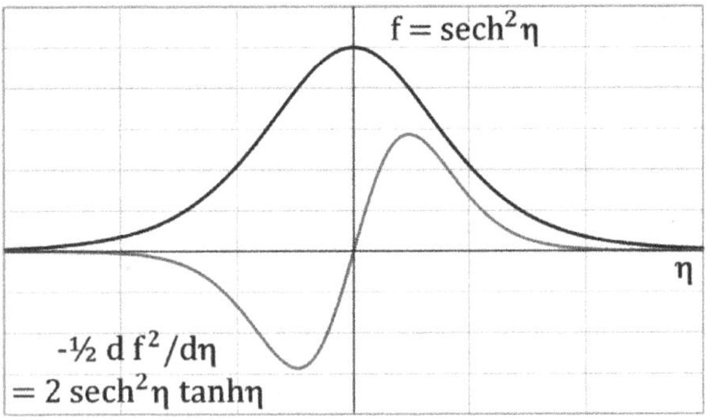

Figure 7.4. Ion acoustic soliton and the self-induced ponderomotive force to balance the dispersion.

7.2.4 *Soliton solution with Bäcklund transform*

Bäcklund transform is a technique for non-linear superposition. It is applied to derive the soliton solutions of the K-dV equation (6.10). A function u with the property $u_\eta = \phi$ is introduced, where $\mathrm{u}_\eta = \frac{\partial}{\partial\eta}\mathrm{u}$. This function is then substituted into (6.10) to convert it to

$$\frac{\partial}{\partial\tau}\mathrm{u} + \frac{\partial^3}{\partial\eta^3}\mathrm{u} + 3\mathrm{u}_\eta^2 = 0 \tag{7.14}$$

Consequently, (7.14) can be decomposed into a set of equations, representing a Bäcklund transform, as follows:

$$\mathrm{u}_\eta = \beta - \frac{1}{2}\mathrm{u}^2 \tag{7.15a}$$

$$\mathrm{u}_\tau = \mathrm{u}\mathrm{u}_{\eta\eta} - 2\mathrm{u}_\eta^2 \tag{7.15b}$$

where β is called Bäcklund parameter. Set $u = \sqrt{2\beta}\tanh\theta(\eta,\tau)$ and substitute it into (7.15a) and (7.15b), yield

$$\theta_\eta = \sqrt{\frac{\beta}{2}} \quad \text{and} \quad \theta_\tau = -2\beta\sqrt{\frac{\beta}{2}}$$

Thus, $\theta(\eta,\tau) = \sqrt{\frac{\beta}{2}}(\eta - 2\beta\tau)$ and

$$\mathrm{u} = \sqrt{2\beta}\tanh\sqrt{\frac{\beta}{2}}(\eta - 2\beta\tau) \tag{7.15c}$$

A soliton solution of (6.10) is then obtained to be

$$\phi = u_\eta = \beta \text{sech}^2 \sqrt{\frac{\beta}{2}}(\eta - 2\beta\tau), \tag{7.16}$$

which is the same as (7.13) with $\beta = \frac{A_s}{2}$.

7.2.5 *Transition from nonstationary to stationary*

For the initial value problem, if the initial wave is a nonstationary nonlinear wave, the wave will steepen and break at the discontinuity during the propagation; then on leaving the transition region, it decomposes into different types of stationary waves. The determination of the wave behavior in the transition region at the discontinuity as well as after becoming stationary waves is a major interest. An analytical approach to study the initial value problem is the inverse scattering method. In a quantum mechanical system, the potential function could be reconstructed via the scattering data.

A. Inverse scattering transform (IST)

It was established in the 1950s that a localized potential field $V(x)$ of the Schrödinger equation

$$\varphi_{xx} + [E - V(x)]\varphi = 0 \tag{7.17}$$

can be completely reconstructed from the scattering data S. If $V(x)$ is in the form of a potential well, then there will be a finite number of bound states with discrete eigen-energies $E_n = -\kappa_n^2$; n $= 1, 2, \ldots$ N, and a continuum of states with eigen-energies $E = k^2$. These eigenvalues determine the corresponding eigen-functions asymptotically to be

1. For the bound states

$$\varphi_n \sim c_n e^{-\kappa_n x}\, as\, x \to +\infty$$

and (7.17a)

$$\varphi_n \propto e^{\kappa_n x}\, as\, x \to -\infty$$

where c_n is determined via the normalization: $\int_{-\infty}^{\infty} \varphi_n^2\, dx = 1$.

2. For the unbound states

$$\varphi \sim e^{-ikx} + b(k)e^{ikx} \; as \, x \to \infty$$

and (7.17b)

$$\varphi \sim a(k)e^{-ikx} \; as \, x \to -\infty.$$

where b(k) represents a reflection coefficient and a(k) a transmission coefficient.

The results render the scattering data $S = \{\kappa_n, c_n;\, a(k),\, b(k)\}$. The corresponding inverse scattering transform (IST) mapping $S \to V$ is accomplished through the Gelfand - Levitan - Marchenko (GLM) linear integral equation. Define the function

$$F(x) = \sum_{n=1}^{N} c_n^2 e^{-\kappa_n x} + \frac{1}{2\pi} \int_{-\infty}^{\infty} b(k) e^{ikx}\, dk, \tag{7.18}$$

then, the potential $V(x)$ is restored from the formula

$$V(x) = -2\frac{\partial}{\partial x} K(x,\, y = x), \tag{7.19}$$

where $K(x, y)$ is found from the (GLM) linear integral equation

$$K(x,y) + F(x+y) + \int_{x}^{\infty} K(x,z)F(y+z)dz = 0. \tag{7.20}$$

If the solution $\phi(\eta, \tau)$ of the K-dV equation (6.10) can be treated as a potential function of a quantum system, where τ is treated as a parameter, i.e., $\phi(\eta, \tau) = V(\eta; \tau)$, then the problem of solving a nonlinear partial differential equation (K-dV equation) in η and τ is reduced to solving two linear equations, a linear SchrodingerSchrödinger equation to determine the scattering data and the Gelfand-Levitan-Marchenko integral equation to obtain the potential function from the scattering data.

Use the relation of the Miura transformation (7.9), in which $\phi(\eta, \tau)$ is the solution of the K-dV equation (6.10), and apply to (7.9) the Cole-Hopf transformation:

$$U = \frac{\psi_\eta}{\psi},$$

a linear Schrödinger equation is derived to be

$$\psi_{\eta\eta} + \phi\psi = 0, \tag{7.21}$$

in which

$$-\phi(\eta, \tau) = V(\eta; \tau) = -2\frac{\partial}{\partial \eta}K(\eta, \eta; \tau) \tag{7.19a}$$

represents the potential function of a quantum system with τ treated as a parameter. In other words, $\phi(\eta, \tau) = \phi(\eta; \tau)$ is used to define a quantum system, which is descriptive by the Schrödinger eigenvalue equation

$$\psi_{\eta\eta} + \phi\psi = -E\psi \tag{7.22a}$$

which has the same form as (7.17), but ϕ and ψ depend parametrically on τ. The initial condition $\phi(\eta, 0)$ determines the initial scattering data $S(0) = \{\kappa_n, c_n(0);\ a(k; 0), b(k; 0)\}$.

Next, introduce the relation

$$\psi_\tau = (\phi_\eta + \gamma)\psi + (4E - 2\phi)\psi_\eta \tag{7.22b}$$

where γ is a constant proportional to the eigenvalue E. For the bound states, $\psi_\tau, \phi_\eta, and\ \phi \sim 0, and\ \frac{\psi_\eta}{\psi} \sim \kappa_n, as\ x \to -\infty$; thus, $\gamma_n = 4\kappa_n^3$. (7.22a) and (7.22b) are the Lax pair which synthesize the K-dV equation, as can be shown that

$$\psi_{\eta\eta\tau} - \psi_{\tau\eta\eta} = -(\phi_\tau + 6\phi\phi_\eta + \phi_{\eta\eta\eta})\psi = 0$$

Equation (7.22a) defines the forward scattering problem. (7.22b) defines the time evolution of the scattering; it updates the scattering data used in setting up the Gelfand-Levitan- Marchenko integral equation.

With the aid of (7.22b), the normalization of the eigenfunctions, and the asymptotic properties of $\psi\,(i.e., \varphi\, in\,(7.17))$, the time evolution of the scattering data $c_n(0)$ of the bound states is determined as follows:

The normalization of the eigenfunction:

$$\int_{-\infty}^{\infty} \psi_n^2\, d\eta = 1 = c_n^{-2} \int_{-\infty}^{\infty} \psi^2\, d\eta$$

leads to the rate equation for c_n^2 as

$$\frac{d}{d\tau}c_n^2 = 2\int_{-\infty}^{\infty}\psi\psi_\tau\,d\eta = 2\int_{-\infty}^{\infty}\psi[(\phi_\eta+\gamma_n)\psi+(4E-2\phi)\psi_\eta]d\eta$$

$$= 2\int_{-\infty}^{\infty}[(\psi^2\phi)_\eta+4(\psi_{\eta\eta}+2E\psi)\psi_\eta+\gamma_n\psi^2]d\eta = 2\gamma_n c_n^2$$

Thus, $c_n^2(\tau) = c_n^2(0)e^{8\kappa_n^3\tau}$.

Next, substitute the asymptotic properties (7.17b) of ψ for the unbound states into (7.22), and note that $\phi\, and\, \phi_\eta \to 0\, as\, |\eta| \to \infty$, it leads to

$$E = k^2 \quad \text{and} \quad \gamma = i4kE = i4k^3; \quad \text{and}$$

$$\frac{d}{d\tau}a(k;\,\tau) = 0 \quad \text{and} \quad \frac{d}{d\tau}b(k;\,\tau) = 8ik^3b(k;\,\tau)$$

Thus $a(k,\tau) = a(k,\,0)\; and\; b(k,\tau) = b(k,\,0)e^{8ik^3\tau}$.

Apply the scattering data $S(\tau)=\{\kappa_n, c_n(0)e^{4\kappa_n^3\tau}; a(k;0), b(k;0)e^{8ik^3\tau}\}$, (7.18) to (7.20) are solved to obtain the potential field $\phi(\eta,\tau)$ of (7.22a).

If reflectionless potential is assumed, i.e., $b(k,\,0) = 0$, it turns out that the potential is a sum of solitary functions ($\propto$ sech^2). If the pulse is not a soliton initially, as time proceeds, the pulse will become individual soliton solutions. Strong non-soliton pulses will break into a train of soliton solutions. The number of solitons equals to the number of bound states in the initial potential well. Apply inverse scattering transforms, N-soliton solutions can be obtained.

B. Example, a two-soliton solution

Consider the initial condition: $\phi(x,\,0) = 6\,\text{sech}^2(x)$, which represents a localized potential well of the Schrödinger equation (7.22a). We first solve (7.22a) to find the eigenvalues and eigenfunctions. This is done by introducing the transform s = tanh(x), which maps x from $(-\infty,\,\infty)$ to s from $(-1,1)$. With the aid of

$$\frac{d^2}{dx^2} \to (1-s^2)^2\frac{d^2}{ds^2} - 2s(1-s^2)\frac{d}{ds} = (1-s^2)\frac{d}{ds}(1-s^2)\frac{d}{ds}$$

and set $U(s, 0) = \psi(x, 0)$, (7.22a) is transformed to an associated Legendre differential equation

$$\frac{d}{ds}(1-s^2)\frac{d}{ds}U + \left[\ell(\ell+1) - \frac{m^2}{(1-s^2)}\right]U = 0 \qquad (7.23)$$

where $\ell = 2$ and $m^2 = -E$. Because $0 < |m| \leq \ell = 2$, $m = 1$ and 2; it leads to two eigenvalues $E_1 = -1$ *and* $E_2 = -4$, giving $\kappa_1 = 1$ *and* $\kappa_2 = 2$. The two eigenfunctions are the associated Legendre polynomials P_2^1 *and* P_2^2, which are

$$P_2^1 = -\frac{1}{2}(1-s^2)^{\frac{1}{2}}\frac{d}{ds}(3s^2-1) = -3s(1-s^2)^{\frac{1}{2}}$$

and

$$P_2^2 = \frac{1}{2}(1-s^2)\frac{d^2}{ds^2}(3s^2-1) = 3(1-s^2)$$

Thus, normalized eigenfunctions are obtained to be

$$\varphi_1(x,0) = -\sqrt{\frac{3}{2}}\,\mathrm{tanh(x)\,sech(x)} \quad \text{and} \quad \varphi_2(x,0) = \frac{\sqrt{3}}{2}\,\mathrm{sech^2(x)}$$

Apply the asymptotic relation (7.17a), yields

$$c_1(0) = \lim_{x\to\infty}\varphi_1(x,0)e^{\kappa_1 x} = \lim_{x\to\infty}[-\sqrt{3/2}\,tanh(x)\,sech(x)e^x] = -\sqrt{6}$$

$$c_2(0) = \lim_{x\to\infty}\varphi_2(x,0)e^{\kappa_2 x} = \lim_{x\to\infty}[(\sqrt{3}/2)\,sech^2(x)e^{2x}] = 2\sqrt{3}$$

Because soliton solutions are only related to the bound states, (7.18) is determined explicitly to be

$$\begin{aligned} F(x;t) &= \sum_{n=1}^{2}[c_n^2(0)e^{8\kappa_n^3 t}e^{-\kappa_n x}] \\ &= (6\,e^{8t-x} + 12\,e^{64t-2x}) \end{aligned} \qquad (7.18a)$$

Substitute (7.18a) into (7.20), the GLM linear integral equation becomes

$$\begin{aligned} &K(x,\,y;t) + 6\,e^{8t-(x+y)} + 12\,e^{64t-2(x+y)} \\ &\quad + \int_x^\infty K(x,z;t)[6\,e^{8t-(y+z)} + 12\,e^{64t-2(y+z)}]dz = 0. \end{aligned} \qquad (7.20a)$$

Eq. (7.20a) is solved via the separation of variables. Substitute

$$K(x, y; t) = H_1(x; t)e^{-y} + H_2(x; t)e^{-2y}$$

into (7.20a) and group the terms having the same y-dependent exponential factor (*i.e.*, $e^{-y}\, and\, e^{-2y}$ groups), two equations for $H_1\, and\, H_2$ are derived to be

$$H_1(x; t) + 6\,e^{8t-x} + 6\int_x^\infty [H_1(x; t)e^{-z} + H_2(x; t)e^{-2z}]\, e^{8t-z}dz = 0$$

$$H_2(x; t) + 12\,e^{64t-2x} + 12\int_x^\infty [H_1(x; t)e^{-z} + H_2(x; t)e^{-2z}]\, e^{64t-2z}dz = 0.$$

which lead to two coupled algebraic equations

$$(1 + 3e^{8t-2x})H_1 + 2e^{8t-3x}H_2 = -6\,e^{8t-x} \tag{7.24a}$$

and

$$4e^{64t-3x}H_1 + (1 + 3e^{64t-4x})H_2 = -12\,e^{64t-2x} \tag{7.24b}$$

These two equations are solved, for H_1 and H_2 to obtain

$$K(x, x; t) = -6\frac{(e^{8t-2x} + e^{72t-6x} + 2e^{64t-4x})}{1 + 3e^{64t-4x} + 3e^{8t-2x} + e^{72t-6x}}$$

Substitute $K(x, x; t)$ into (7.19a), a two-soliton solution is obtained to be:

$$\phi(x, t) = 12\left(\frac{3 + 4\cosh(2x - 8t) + \cosh(4x - 64t)}{[3\cosh(x - 28t) + \cosh(3x - 36t)]^2}\right) \tag{7.25}$$

It satisfies the initial condition: $\phi(x,\, 0) = 6\,\text{sech}^2(x)$

C. Asymptotic form of the two-soliton solution

One of the unique properties of solitons is that they can interact with other solitons, and emerge unchanged after separation, except for a phase shift. Hence, as t $\to \infty$, $\phi(x, t)$ should emerge as a sum of two separate solitons, i.e., the two hyperbolic cosine terms in the numerator of (7.25) are converted to the forms given by (7.16); in other words, (7.25) has the asymptotic form

$$\phi(x, t \to \infty) = 8\,\mathrm{sech}^2[2(x - 16t) + \theta_1] + 2\,\mathrm{sech}^2[(\mathrm{x} - 4\mathrm{t}) + \theta_2] \tag{7.25a}$$

To verify (7.25a), we introduce

$$y = \mathrm{x} - 16\mathrm{t} \text{ and } \mathrm{z} = \mathrm{x} - 4\mathrm{t}$$

and rewrite

$$2x - 8t = 3y - z + 36t;$$

$$4x - 64t = y + 3z - 36t;$$

$$x - 28t = \begin{cases} y - 12t \\ z - 24t \end{cases};$$

$$3x - 36t = \begin{cases} 3y + 12t \\ 4y - z + 24t \end{cases} = \begin{cases} y + 2z - 12t \\ 3z - 24t \end{cases};$$

$$3\cosh(x - 28t) + \cosh(3x - 36t) = \begin{cases} A \\ B \\ C \\ D \end{cases}$$

where

$$A = 3\cosh(y - 12t) + \cosh(3y + 12t)$$

$$B = 3\cosh(z - 24t) + \cosh(4y - z + 24t)$$

$$C = 3\cosh(y - 12t) + \cosh(y + 2z - 12t)$$

$$D = 3\cosh(z - 24t) + \cosh(3z - 24t)$$

thus, (7.25) becomes

$$\phi(x, t) = \frac{36}{[3\cosh(y - 12t) + \cosh(3z - 24t)]^2} + \frac{48\cosh(3y - z + 36t)}{AB} + \frac{12\cosh(y + 3z - 36t)}{CD} \tag{7.25b}$$

Keep y and z finite, as t $\to \infty$, then

$$\phi(x, t \to \infty) = \frac{96e^{3y-z}}{9e^{-y-z} + 3e^{3y-z} + 3e^{3y-z} + e^{7y-z}} + \frac{24e^{-y-3z}}{9e^{-y-z} + 3e^{-y-3z} + 3e^{-y-3z} + e^{-y-5z}}$$

$$= \frac{96}{9e^{-4y} + 6 + e^{4y}} + \frac{24}{9e^{2z} + 6 + e^{-2z}}$$

$$= \frac{32}{[e^{(2y-ln\sqrt{3})} + e^{-(2y-ln\sqrt{3})}]^2} + \frac{8}{[e^{(z+ln\sqrt{3})} + e^{-(z+ln\sqrt{3})}]^2}$$

$$= 8\,sech^2(2y - ln\sqrt{3}) + 2\,sech^2(z + ln\sqrt{3}), \qquad (7.25c)$$

which is the same as (7.25a) with $\theta_{1,2} = \mp ln\sqrt{3}$. (7.25) can be plotted to observe the structure of the two-soliton solution of the K-dV equation during the interaction.

D. Pulse behavior in the transition region

A numerical solution of the Korteweg-deVries equation (Sec. 6.2.2)

$$\frac{\partial\phi}{\partial t} + \phi\frac{\partial\phi}{\partial x} + 0.001\frac{\partial^3\phi}{\partial x^3} = 0$$

with the initial condition $\phi(x, 0) = \exp[-(5x/3)^2]$, is presented in Fig. 7.5. It shows that an initial Gaussian pulse is decomposed into multiple solitons.

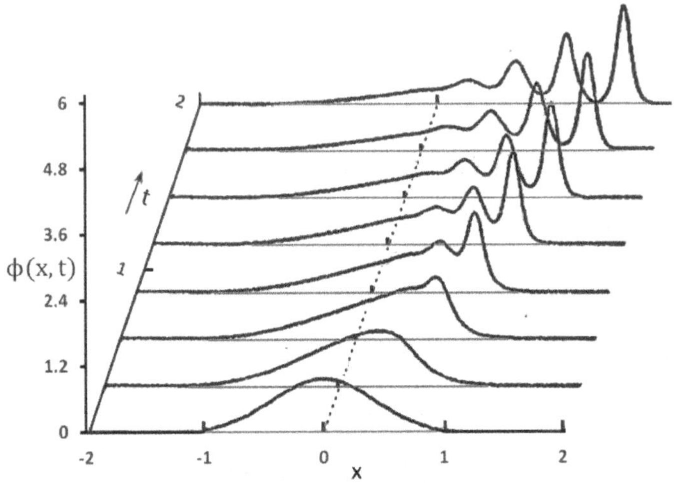

Figure 7.5. Numerical solution of Korteweg-de Vries equation.

It also shows that larger amplitude one is narrower and moves faster as indicated by (7.16); their separation increases with time. The transition process illustrates the complexity of such nonlinear wave behavior.

7.3 Burgers equation

Burgers equation describes nonlinear wave motion undergone linear diffusion and is the simplest model for analyzing combined effect of nonlinear advection and diffusion. On the LHS of (6.12), the first term characterizes the time evolution of the wave propagating in one direction and the second term induces nonlinear convection; the RHS term induces linear diffusion.

The nonlinear nature of Burgers equation has been exploited as a useful prototype differential equation for modeling many phenomena such as shock flows, wave propagation in combustion chambers, vehicular traffic movement, acoustic transmission, etc. It is applied to study flow phenomenon regarding to the balancing effects of viscous and inertial or convective forces. When inertia or convective forces are dominant, its solution resembles that of the kinematic wave equation which displays a propagating wave front and boundary layers. In contrast, when viscous forces are dominant, propagating wave front is smeared and diffused due to viscous action.

7.3.1 *Analytical solution via the Cole-Hopf transformation*

Introduce new notations: $\mathrm{u}(\mathrm{x},\mathrm{t}) = \phi_1(\eta,\tau)$, where $\mathrm{x} = \eta$ and $\mathrm{t} = \tau$, and $u_t = \frac{\partial}{\partial t}u;\, u_x = \frac{\partial}{\partial x}u;\, and\, u_{xx} = \frac{\partial^2}{\partial x^2}u$, (6.12) is re-expressed as

$$u_t + uu_x = bu_{xx} \tag{6.12a}$$

The initial value problem for the viscid Burgers equation (6.12a) is solved analytically by introducing the Cole-Hopf transformation:

$$u = -2b\frac{\varphi_x}{\varphi} = -2b\frac{\partial}{\partial x}\ln\varphi \tag{7.26}$$

Operating on (7.26) for the terms in (6.12a), yields

$$\mathrm{u_t} = 2\mathrm{b}\frac{\varphi_\mathrm{t}\varphi_\mathrm{x} - \varphi\varphi_\mathrm{tx}}{\varphi^2},\ \mathrm{uu_x} = 4\mathrm{b}^2\frac{\varphi_\mathrm{x}(\varphi\varphi_\mathrm{xx} - \varphi_\mathrm{x}^2)}{\varphi^3} \text{ and}$$

$$bu_{xx} = -2b^2\frac{\varphi^2\varphi_{xxx} - 3\varphi\varphi_x\varphi_{xx} + 2\varphi_x^3}{\varphi^3}$$

With the aid of these relations, (6.12a) becomes

$$2b\frac{-\varphi\varphi_{\mathrm{tx}}+\varphi_{\mathrm{x}}\left(\varphi_t-\mathrm{b}\varphi_{\mathrm{xx}}\right)+b\varphi\varphi_{xxx}}{\varphi^2}=0 \tag{7.27a}$$

It leads to

$$\varphi_x\left(\varphi_t-b\varphi_{xx}\right)=\varphi(\varphi_{tx}-b\varphi_{xxx})=\varphi(\varphi_t-b\varphi_{xx})_x \tag{7.27b}$$

Eq. (7.27b) shows that φ is governed by a diffusion equation

$$\varphi_{\mathrm{t}}-b\varphi_{\mathrm{xx}}=0, \tag{7.28}$$

where b is the diffusion coefficient. This equation was considered in Chapter1 with D = ∓b and E = 0 set in (1.12b). Its solution for an initial Gaussian pulse (1.15) is given by (1.12c). The general solution of the diffusion equation is also well known, however, in the current case, its initial condition $\varphi(x,\,0)=\varphi_0(x)$ is imposed by the initial condition $u(x,\,0)=u_0(x)$ of the Burgers equation (6.12a) and can be determined via the transformation (7.26) that is integrated to be

$$\varphi(x,t)=exp\left(-\frac{1}{2b}\int_0^x u(\xi,t)d\xi\right) \tag{7.29}$$

Then the initial condition of the diffusion equation is given to be

$$\varphi_0(x)=exp\left(-\frac{1}{2b}\int_0^x u_0(\xi)d\xi\right). \tag{7.29a}$$

The diffusion equation (7.28) is solved by taking the Fourier transform with respect to x and then integrating on t; it yields

$$\hat{\psi}(k,t)=\hat{\psi}_0(k)e^{-bk^2t} \tag{7.30}$$

where the Fourier transform is defined to be

$$\hat{\psi}(k,t)=\int_{-\infty}^{\infty}\varphi(x,t)e^{-ikx}\,dx.$$

The solution $\varphi(x,t)$ of (7.28) is the inverse Fourier transform of (7.30). Because the RHS of (7.30) is a product of two transform functions, the inverse Fourier transform of (7.30) is the convolution of those two functions in the x space. The inverse Fourier transform of e^{-bk^2t} is evaluated to be

$$\int_{-\infty}^{\infty} e^{-bk^2t} e^{ikx} \frac{dk}{2\pi} = \frac{1}{2\sqrt{\pi bt}} e^{-\frac{x^2}{4bt}} = g(x,t)$$

Then,

$$\begin{aligned}\varphi(x,t) &= \int_{-\infty}^{\infty} \hat{\psi}(k,t) e^{ikx} \frac{dk}{2\pi} = \varphi_0(x) \circ g(x,t) \\ &= \frac{1}{2\sqrt{\pi bt}} \int_{-\infty}^{\infty} \varphi_0(\xi) e^{-\frac{(x-\xi)^2}{4bt}}\, d\xi \end{aligned} \tag{7.31}$$

where "o" denotes the convolution product. Substitute (7.31) into (7.26), the analytical solution of (6.12a) is obtained to be

$$u(x,\, t) = \frac{\int_{-\infty}^{\infty} \frac{(x-\xi)}{t} \varphi_0(\xi) e^{-\frac{(x-\xi)^2}{4bt}}\, d\xi}{\int_{-\infty}^{\infty} \varphi_0(\xi) e^{-\frac{(x-\xi)^2}{4bt}}\, d\xi} \tag{7.32}$$

7.3.2 *Propagating modes*

Consider traveling wave solution of the form $\phi_1(\eta,\tau) = \mathrm{F}_1(\xi)$, where $\xi = \eta - A_b\tau$, (6.12) becomes

$$\mathrm{b}\frac{\mathrm{d}^2}{\mathrm{d}\xi^2}\mathrm{F}_1 + \frac{\mathrm{d}}{\mathrm{d}\xi}\left(A_\mathrm{b}\mathrm{F}_1 - \frac{1}{2}\mathrm{F}_1^2\right) = 0 \tag{7.33}$$

It is integrated to be

$$\mathrm{b}\frac{\mathrm{d}}{\mathrm{d}\xi}\mathrm{F}_1 + \left(A_\mathrm{b}\mathrm{F}_1 - \frac{1}{2}\mathrm{F}_1^2\right) = 0 \tag{7.33a}$$

where the integration constant is set to zero for the same reason that it can be cancelled by a constant shift of F_1. Eq. (7.33a) is integrated to obtain

$$\mathrm{F}_1(\xi) = A_\mathrm{b}\left(1 - \tanh\frac{A_\mathrm{b}}{2\mathrm{b}}\xi\right) \tag{7.34}$$

It shows that F_1 starts at a level of $2A_\mathrm{b}$, i.e., $\mathrm{F}_1(-\infty) = 2A_\mathrm{b}$, and decreases continuously to zero as it moves to ∞, i.e., $\mathrm{F}_1(\infty) = 0$. The transition width is proportional to $2\mathrm{b}/A_\mathrm{b}$, i.e., proportional to the damping factor b and inversely proportional to the amplitude A_b. As $\mathrm{b} \to 0$, it becomes a shock wave as shown in Fig. 6.1d. The shock front has a step transition at $\xi = 0$.

Problems

P7.1. In linear system, the Schrödinger eigenvalue equation (7.2) becomes

$$\left(-{}^{1}\!/\!_{2}\frac{d^2}{d\xi^2} + \mathrm{V}(\xi)\right)\phi = \mathrm{E}\phi \tag{P7.1}$$

where $\mathrm{V}(\xi)$ is the potential function of the system. Assume that $\mathrm{V}(\xi) = -\mathrm{k}^2\,\mathrm{sech}^2\,(\mathrm{kx})$, find the solitary eigen-function $\phi(\xi)$ and the corresponding eigenvalue E.

P7.2. In problem P7.1., change the potential function to $\mathrm{V}(\xi) = -\exp(-2|\xi|)$; show that with the transform $\mathrm{z} = \sqrt{2}\mathrm{e}^{\xi}$ for $\xi < 0$ and $\mathrm{z} = \sqrt{2}\mathrm{e}^{-\xi}$ for $\xi > 0$, the function $\mathrm{y(z)} = \phi(\xi)$ satisfies a Bessel equation, i.e., (P7.1) is transformed to the Bessel differential equation

$$\frac{d^2}{d\mathrm{z}^2}\mathrm{y} + \frac{1}{\mathrm{z}}\frac{d}{d\mathrm{z}}\mathrm{y} + \left(1 - \frac{\nu^2}{\mathrm{z}^2}\right)\mathrm{y} = 0$$

where $\nu^2 = -2\mathrm{E}$.

P7.3. Show that the Hamiltonian H in (7.1c) is a constant of motion.

P7.4. The sine-Gordon equation

$$u_{tt} - u_{xx} + \sin u = 0 \tag{P7.2}$$

is a nonlinear hyperbolic partial differential equation, where $u = u(x,t)$. In the low-amplitude case ($\sin u \approx u$), it reduces to the Klein-Gordon equation

$$u_{tt} - u_{xx} + u = 0$$

Find a general plane wave solution.

P7.5. Find the travelling wave solution of the sine-Gordon equation (P7.2).

P7.6. The initial condition leads to a pure 1-soliton solution for the K-dV equation (6.10) is given to be $\phi(\eta,0) = \frac{\mathrm{A}}{2}\mathrm{sech}^2(\frac{1}{2}\sqrt{\mathrm{A}}\eta)$ Apply IST to solve (6.10) for the initial condition: $\phi(\eta,0) = 2\mathrm{sech}^2(\eta)$

P7.7. In problem P7.6, the initial condition is changed to $\phi(\eta,0) = 24\,\mathrm{sech}^2 2\eta$, find the solution $\phi(\eta,\tau)$ to the K-dV equation (6.10).

Chapter 8

Wave-wave and Wave-particle Interactions

Nonlinear and turbulent wave phenomena occur in the propagation of waves through "nonlinear" media, such as plasmas. Nonlinearities and turbulence may occur separately at low wave intensity, but at high intensity levels they frequently occur simultaneously. In the latter case, there are usually wave dependent changes in the material media supporting the waves, such as the index of refraction of the medium. In plasmas, the modifications include the density and velocity distributions of the charged particles in plasmas. In the previous chapter, nonlinear equations modeling wave propagation in plasma with wave-induced background changes, are derived. The solution of each equation represents a coherent nonlinear wave.

When many waves are excited and propagate simultaneously in a nonlinear medium, those waves can couple together through the nonlinearities of the medium. It results to the generation of new waves; the coupling can be particularly strong when certain matching conditions, such as the frequency and wavevector matching conditions for a new mode, are met. The coupling process widens the spectra of the waves; as the fluctuation levels of those wave are intensified, medium evolves to a turbulent state. In this chapter, wave-wave and wave-particle interactions in turbulent plasma is studied via Vlasov-Poisson equations.

8.1 Vlasov-Poisson system

The phase space density distribution function f_a of a plasma species is governed by the Vlasov equation

$$\frac{\partial f_a}{\partial t} + \mathbf{v} \cdot \nabla f_a + \mathbf{E} \cdot \nabla_{\mathbf{v}} f_a = 0 \tag{8.1a}$$

This equation represents a continuity equation or a charge particle conservation equation, where **E** is electrostatic wave field governed by the Poisson equation

$$\nabla \cdot \mathbf{E} = \frac{\rho}{\epsilon_0} = e \int (f_i - f_e)\, d\mathbf{v} = e(n_i - n_e) \tag{8.1b}$$

where ρ is the induced charge density by the wave electric field **E** in plasma; singly charged ions are assumed; ϵ_0 is the free space permittivity.

At the linear level, i.e., with small wave field E, (8.1a) is linearized with respect to the first order perturbation field **E** by setting $f_a = f_{a0} + (f_{a1} + c.c.)$, where "c.c." represents complex conjugate; f_{a0} is the unperturbed background distribution and f_{a1} is the first order response to the wave field E. Substitute $f_a = f_{a0} + (f_{a1} + c.c.)$ into (8.1a), the zeroth and first order equations are obtained respectively to be

$$\frac{\partial f_{a0}}{\partial t} + \mathbf{v} \cdot \nabla f_{a0} = 0 \tag{8.2a}$$

and

$$\frac{\partial f_{a1}}{\partial t} + \mathbf{v} \cdot \nabla f_{a1} = -\left(\frac{q_a}{m_a}\right) \mathbf{E} \cdot \nabla_{\mathbf{v}}\, f_{a0} \tag{8.2b}$$

The solution of (8.2a) is often a Maxwellian,

$$f_{a0} = n_0 (m_a/2\pi T_a)^{3/2} \exp(-m_a v^2/2T_a), \tag{8.2c}$$

In the analysis for a linear system, Eq. 8.2b, a linearization of (8.1a), is employed; it is combined with the Poisson equation (8.1b), to study the linear electrostatic wave properties of uniform unmagnetized plasma (i.e., substitute the solutions of (8.2b) into (8.1b) to determine the relations for nontrivial solutions; those relations, called "dispersion equations" lead to the linear dispersion relations of the plasma waves). Superposition principle is applicable in linear systems; thus, each linear wave is a time-harmonic wave having a function form of $\mathbf{E}_k + \text{c.c.} = \tilde{\boldsymbol{E}}_k e^{i(\mathbf{k}\cdot\mathbf{r}-\omega t)} + \text{c.c.}$, where $\tilde{E}_k$, $\mathbf{k}$, and ω are the amplitude, and the wavevector and frequency of the wave; and relations in the expression of $\mathrm{k} = \mathrm{k}(\omega)$ or $\omega = \omega(\mathrm{k})$, deduced from the conditions for non-trivial solutions of the coupled system equations (8.1b) and (8.2b), are called the linear dispersion relations of the plasma waves.

8.2 Velocity diffusion

When the wave amplitude $\tilde{E}_k$ of a single harmonic wave increases, it can trap charged particles, which move at close to the wave velocity, to flatten the background velocity distribution in the resonance region (i.e., around the phase velocity of the wave). This process is described by a diffusion equation, which is derived by including a filtered second order term to (8.2a), which is updated to be

$$\frac{\partial \mathrm{f}_{a0}}{\partial t} + \mathbf{v}\cdot\nabla \mathrm{f}_{a0} = -2\left(\frac{\mathrm{q}_a}{\mathrm{m}_a}\right) R_e[\langle \boldsymbol{E}_k^* \cdot \nabla_{\mathbf{v}} \mathrm{f}_{a1}\rangle] \tag{8.3a}$$

where R_e stands for real part and the notation $\langle\cdots\rangle$ denotes a low-pass filter. Eq. (8.2b), applying the method of characteristics, is integrated to obtain

$$\mathrm{f}_{a1} = -i\left(\frac{\mathrm{q}_a}{\mathrm{m}_a}\right)\frac{\mathbf{E}_k\cdot\nabla_{\mathbf{v}}\mathrm{f}_{a0}}{\omega - \mathbf{k}\cdot\mathbf{v}},$$

which is then substituted into (8.3a); a linear diffusion equation is obtained to be

$$\frac{\partial}{\partial t}\mathrm{f}_{a0} + \mathbf{v}\cdot\nabla \mathrm{f}_{a0} = \nabla_{\mathbf{v}}\cdot \underline{\underline{\mathbf{D}}}_k(\mathbf{v})\cdot\nabla_{\mathbf{v}}\mathrm{f}_{a0} \tag{8.3b}$$

where the diffusion operator

$$\underline{\underline{\mathbf{D}}}_{\mathbf{k}}(\mathbf{v}) = 2\left(\frac{\mathrm{q}_a}{\mathrm{m}_a}\right)^2 \frac{\gamma_k \tilde{\boldsymbol{E}}_k \tilde{\boldsymbol{E}}_{\mathbf{k}}^*}{[(\mathbf{k}\cdot\mathbf{v}-\omega_r)^2+\gamma_k^2]}$$

and $\omega = \omega_r + i\gamma_k$ is set with $\gamma_k \gtrless 0$ for a growing/damped wave.

8.3 Mode coupling

In high energy state, plasma waves are excited in different branches of modes and grow to large amplitudes and wide bands. Those waves couple to each other through the nonlinearity of the plasma to generate new plasma waves. In the following, mode coupling of electrostatic modes in unmagnetized plasma is formulated via a kinetic approach. It also illustrates the approach of applying the Eulerian specification of the variables $(\mathrm{r}, \mathrm{v}, \mathrm{t})$, which are considered to be independent of one another, to analyze the kinetic equation (8.1a).

Expend (8.1a) with respect to the first order perturbation field $\mathbf{E_1} + \mathbf{c.c.}$ further by setting $\mathrm{f_a} = \mathrm{f_{a0}} + (\mathrm{f_{a1}} + \mathrm{f_{a2}} + c.c.)$ and $\mathbf{E} = \mathbf{E_1} + \mathbf{E_2} + \mathbf{c.c.}$, where $\mathrm{f_{a2}}$ is the second order (nonlinear) response to $\mathbf{E_1}$; $\mathbf{E_2}$ is the induced self-consistent (second-order) field responding to $\mathrm{f_{a2}}$.

Substitute $\mathrm{f_a} = \mathrm{f_{a0}} + (\mathrm{f_{a1}} + \mathrm{f_{a2}} + c.c.)$ and $\mathbf{E} = \mathbf{E_1} + \mathbf{E_2} + \mathbf{c.c.}$ into (8.1a), the zeroth, first, and second order equations are obtained respectively to be

$$\frac{\partial \mathrm{f_{a0}}}{\partial \mathrm{t}} + \mathbf{v} \cdot \nabla \mathrm{f_{a0}} = -2\left(\frac{\mathrm{q_a}}{\mathrm{m_a}}\right) R_e[\langle \boldsymbol{E}_1^* \cdot \nabla_{\mathbf{v}} f_{a1} \rangle] \tag{8.4a}$$

$$\frac{\partial \mathrm{f_{a1}}}{\partial \mathrm{t}} + \mathbf{v} \cdot \nabla \mathrm{f_{a1}} = -\left(\frac{\mathrm{q_a}}{\mathrm{m_a}}\right) \boldsymbol{E}_1 \cdot \nabla_{\mathbf{v}} \mathrm{f_{a0}} \tag{8.4b}$$

and

$$\frac{\partial \mathrm{f_{a2}}}{\partial \mathrm{t}} + \mathbf{v} \cdot \nabla \mathrm{f_{a2}} = -\left(\frac{\mathrm{q_a}}{\mathrm{m_a}}\right) \boldsymbol{E}_2 \cdot \nabla_{\mathbf{v}} \mathrm{f_{a0}} - \left(\frac{\mathrm{q_a}}{\mathrm{m_a}}\right) \boldsymbol{E}_1 \cdot \nabla_{\mathbf{v}} \mathrm{f_{a1}} \tag{8.4c}$$

Neglect the second order term on the RHS of (8.4a), in thermal equilibrium the solution of (8.4a) is a Maxwellian, given by (8.2c).

In the presence of an electric field disturbance $\boldsymbol{E}_1(\mathbf{r}, t) = \tilde{E}_{1\mathrm{k}}\exp[i(\mathbf{k} \cdot \mathbf{r} - \omega \mathrm{t})] + \mathrm{c.c.}$, the linear response in (8.4b) is preset to be $\mathrm{f_{a1}}\ (\mathbf{r}, \mathbf{v}, t) = \mathrm{F_{a1}}(\mathbf{v})\exp[i(\mathbf{k} \cdot \mathbf{r} - \omega \mathrm{t})] + \mathrm{c.c.}$, having a similar space-time variation, where $\mathrm{F_{a1}}(\mathbf{v})$ is determined, by substituting the preset solution into (8.4b), to be

$$\mathrm{F_{a1}}(\mathrm{v}) = \mathrm{i}\left(\frac{\mathrm{q_a}}{\mathrm{m_a}}\right) \frac{\tilde{\boldsymbol{E}}_{1\mathrm{k}} \cdot \nabla_{\mathbf{v}} \mathrm{f_{a0}}}{(\mathbf{k} \cdot \mathbf{v} - \omega)} \tag{8.5a}$$

From Gauss's law,

$$\mathbf{ik} \cdot \boldsymbol{E}_1 = \rho_1/\epsilon_0 = \sum_{\mathrm{a=e,i}} \mathrm{q_a} \int \mathrm{f_{a1}}\, \mathrm{dv},$$

it reduces to

$$\begin{aligned} ik\tilde{E}_{1k} &= i \sum_{a=e,i} \frac{\omega_{\mathrm{pa}}^2}{\mathrm{n_0}} \left(\frac{\tilde{E}_{1k}}{\mathrm{k}}\right) \int \frac{\partial_{\mathrm{v_z}} \mathrm{f_{a0}}}{\mathrm{v_z} - \frac{\omega}{\mathrm{k}}} \mathrm{dv} \\ &= -i\mathrm{k} \sum_{a=e,i} \chi_a(\mathrm{k}, \omega) \tilde{E}_{1k} \end{aligned} \tag{8.5b}$$

where $\boldsymbol{E}_1$ is set in the z direction without losing the generality and the electric susceptibility χ_a of the species "a" is given to be

$$\chi_a(k,\omega) = \frac{1}{\sqrt{2\pi}}\left(\frac{\omega_{pa}^2}{k^2 v_{ta}^3}\right)\int \frac{v_z}{v_z - \omega/k}\exp\left(-\frac{m_a v_z^2}{2T_a}\right) dv_z$$

$$= \left(\frac{\omega_{pa}^2}{k^2 v_{ta}^2}\right)[1 + \xi_{a0}\, Z(\xi_{a0})] \tag{8.5c}$$

with $\xi_{a0} = \frac{\omega}{k v_{ta}}$ and

$$Z(\xi_{a0}) = \frac{1}{\sqrt{2\pi}}\int_{-\infty}^{\infty} \frac{e^{-(x^2/2)}}{x - \xi_{a0}}\, dx$$

is the plasma dispersion function. Eq. (8.5b) can be written as

$$ik\left[1 + \sum_{a=e,i}\chi_a(k,\omega)\right]\tilde{E}_{1k} = 0 = ik\epsilon_L(k,\omega)\tilde{E}_{1k},$$

where the relative dielectric function

$$\epsilon_L(k,\omega) = 1 + \sum_{a=e,i}\chi_a(k,\omega)$$

and $\epsilon_L(k,\omega) = 0$ when $\boldsymbol{E}_1(k,\omega)$ is a mode field.

Next, Eq. (8.4c) is applied to determine the fields:

$$\boldsymbol{E}_2^{\pm}(\mathbf{r},t) = \boldsymbol{E}_2^{\pm}(\mathbf{k}_{\pm},\omega_{\pm})\exp[i(\mathbf{k}_{\pm}\cdot\mathbf{r} - \omega_{\pm}t)] + c.c.$$

induced by the coupling of two mode fields:

$$\boldsymbol{E}_{11}(\mathbf{r},t) = \boldsymbol{E}_{11}(\mathbf{k}_1,\omega_1)\exp[i(\mathbf{k}_1\cdot\mathbf{r} - \omega_1 t)] + c.c$$

and

$$\boldsymbol{E}_{12}(\mathbf{r},t) = \boldsymbol{E}_{12}(\mathbf{k}_2,\omega_2)\exp[i(\mathbf{k}_2\cdot\mathbf{r} - \omega_2 t)] + c.c,$$

where

$$\mathbf{k}_{\pm} = \mathbf{k}_1 \pm \mathbf{k}_2 \quad \text{and} \quad \omega_{\pm} = \omega_1 \pm \omega_2.$$

For the concerned fields, (8.4c) becomes

$$\frac{\partial f_{a2}^{\pm}}{\partial t} + \mathbf{v}\cdot\nabla f_{a2}^{\pm} = -\left(\frac{q_a}{m_a}\right)\boldsymbol{E}_2^{\pm}\cdot\nabla_{\mathbf{v}}\, f_{a0}$$

$$-\left(\frac{q_a}{m_a}\right)\langle \boldsymbol{E}_{11}\cdot\nabla_{\mathbf{v}}\, f_{a12} + \boldsymbol{E}_{12}\cdot\nabla_{\mathbf{v}}\, f_{a11}\rangle \tag{8.5d}$$

Substitute $f_{a2}^{\pm} = F_{a2}^{\pm}(v)\exp[i(\mathbf{k}_{\pm}\cdot\mathbf{r} - \omega_{\pm}t)] + c.c.$ into (8.5d), yields

$$
\begin{aligned}
F_{a2}^{\pm}(v) &= i\left(\frac{q_a}{m_a}\right)\frac{(\boldsymbol{E}_2^{\pm}\cdot\nabla_{\mathbf{v}} f_{a0} + \boldsymbol{E}_{11}\cdot\nabla_{\mathbf{v}} F_{a12}^{\pm} + \boldsymbol{E}_{12}^{\pm}\cdot\nabla_{\mathbf{v}} F_{a11})}{\mathbf{k}_{\pm}\cdot\mathbf{v} - \omega_{\pm}} \\
&= i\frac{\left(\frac{q_a}{m_a}\right)}{\mathbf{k}_{\pm}\cdot\mathbf{v} - \omega_{\pm}}\left\{\boldsymbol{E}_2^{\pm}\cdot\nabla_{\mathbf{v}} f_{a0} \pm i\left(\frac{q_a}{m_a}\right)\nabla_{\mathbf{v}}\right. \\
&\quad \left.\cdot\left[\left(\frac{\boldsymbol{E}_{11}\boldsymbol{E}_{12}^{\pm}}{\mathbf{k}_2\cdot\mathbf{v} - \omega_2}\right) \pm \left(\frac{\boldsymbol{E}_{12}^{\pm}\boldsymbol{E}_{11}}{\mathbf{k}_1\cdot\mathbf{v} - \omega_1}\right)\right]\cdot\nabla_{\mathbf{v}} f_{a0}\right\} \qquad (8.5e)
\end{aligned}
$$

where $F_{a12}^{+} = F_{a12}$ and $F_{a12}^{-} = F_{a12}^{*}$; $\boldsymbol{E}_{12}^{+} = \boldsymbol{E}_{12}$ and $\boldsymbol{E}_{12}^{-} = \boldsymbol{E}_{12}^{*}$; and

$$
F_{a11,2}(v) = i\left(\frac{q_a}{m_a}\right)\frac{\boldsymbol{E}_{11,2}\cdot\nabla_{\mathbf{v}} f_{a0}}{\mathbf{k}_{1,2}\cdot\mathbf{v} - \omega_{1,2}},
$$

given by (8.5a), are substituted.

To simplify the formulation, we consider the case of $\boldsymbol{E}_{11}\|\boldsymbol{E}_{12}$, i.e., $\mathbf{k}_1\|\mathbf{k}_2\|\mathbf{k}_{\pm}$, and set them in the z direction, the last two terms on the RHS of (8.5e) can be combined to give

$$
\begin{aligned}
F_{a2}^{\pm}(v) &= i\frac{\left(\frac{q_a}{m_a}\right)}{k_{\pm}v_z - \omega_{\pm}}\left[E_2^{\pm}\partial_{v_z} f_{a0} \pm i\left(\frac{q_a}{m_a}\right)\partial_{v_z}\right. \\
&\quad \left.\times\frac{(k_{\pm}v_z - \omega_{\pm})E_{11}E_{12}^{\pm}}{(k_1 v_z - \omega_1)(k_2 v_z - \omega_2)}\partial_{v_z} f_{a0}\right] \qquad (8.5f)
\end{aligned}
$$

Thus,

$$
\begin{aligned}
ik_{\pm}E_2^{\pm}(k_{\pm},\omega_{\pm}) &= \sum_{a=e,i}\left(\frac{q_a}{\varepsilon_0}\right)\int F_{a2}^{\pm}(v)dv \\
&= -ik_{\pm}\sum_{a=e,i}\left\{\chi_a(k_{\pm},\omega_{\pm})E_2^{\pm}(k_{\pm},\omega_{\pm})\right. \\
&\quad - i\frac{q_a}{m_a}[k_{\pm}^3\chi_a(k_{\pm},\omega_{\pm}) - k_1^3\chi_a(k_1,\omega_1) \mp k_2^3\chi_a(k_2,\omega_2)] \\
&\quad \left.\times\frac{E_{11}E_{12}^{\pm}}{(k_1\omega_2 - k_2\omega_1)^2}\right\} \qquad (8.5g)
\end{aligned}
$$

The second order field induced by the mode coupling is obtained to be

$$E_2^{\pm}(\mathrm{k}_{\pm},\omega_{\pm}) = i\sum_{\mathrm{a=e,i}} \frac{\mathrm{q_a}}{\mathrm{m_a}} \frac{[\mathrm{k}_{\pm}^3\chi_a(\mathrm{k}_{\pm},\omega_{\pm}) - \mathrm{k}_1^3\chi_a(\mathrm{k}_1,\omega_1) \mp \mathrm{k}_2^3\chi_a(\mathrm{k}_2,\omega_2)]}{\epsilon_L(\mathrm{k}_{\pm},\omega_{\pm})} \times \frac{E_{11}E_{12}^{\pm}}{(\mathrm{k}_1\omega_2 - \mathrm{k}_2\omega_1)^2} \tag{8.6}$$

1. Example 1: Evaluate the low frequency fluctuations (k_-,ω_-) induced by large amplitude Langmuir waves.

In this situation, $\chi_e(\mathrm{k}_{1,2},\omega_{1,2}) \cong -1$, $\chi_i(\mathrm{k}_{1,2},\omega_{1,2}) \cong 0$, and $\chi_e(\mathrm{k}_-,\omega_-) \cong k_{De}^2/k_-^2$; thus

$$E_2^-(\mathrm{k}_-,\omega_-) = -i\left(\frac{e}{m_e}\right)\frac{\mathrm{k}_-}{(\mathrm{k}_1\omega_2 - \mathrm{k}_2\omega_1)^2}\left[\frac{k_{De}^2 + k_1^2 + k_1k_2 + k_2^2}{\epsilon_L(k_-,\omega_-)}\right]E_{11}E_{12}^* \tag{8.7}$$

where

$$\omega_- = \omega_1 - \omega_2 = \sqrt{\omega_{\mathrm{pe}}^2 + 3\mathrm{k}_1^2\mathrm{v}_{\mathrm{te}}^2} - \sqrt{\omega_{\mathrm{pe}}^2 + 3\mathrm{k}_2^2\mathrm{v}_{\mathrm{te}}^2} \cong \frac{1.5(\mathrm{k}_1^2 - \mathrm{k}_2^2)\mathrm{v}_{\mathrm{te}}^2}{\omega_{\mathrm{pe}}}.$$

In the coupling of two parallel propagating Langmuir waves, $\omega_- \gg \omega_{\mathrm{sk}} = \mathrm{k}_-\mathrm{C_s}$, $(\mathbf{k}_-,\omega_-)$ is not an ion acoustic mode and $\epsilon_L(\mathrm{k}_-,\omega_-) \cong k_{De}^2/k_-^2 \gg 1$; thus the induced fluctuation field intensity $|E_2^-(\mathrm{k}_-,\omega_-)|$ is low. However, the coupling of two oppositely propagating Langmuir waves $(\mathbf{k}_1,\omega_1)$ and $(\mathbf{k}_2,\omega_2)$, which have slightly different wavenumbers, ion acoustic mode $(\mathbf{k}_\mathrm{s},\omega_\mathrm{s})$ may be generated. It requires that

$$\mathrm{k_s} = \mathrm{k}_- = |\mathrm{k}_1| + |\mathrm{k}_2| \quad \text{and} \quad \omega_\mathrm{s} = \mathrm{k_sC_s} = \frac{1.5(\mathrm{k}_1^2 - \mathrm{k}_2^2)\mathrm{v}_{\mathrm{te}}^2}{\omega_{\mathrm{pe}}},$$

which lead to

$$|\mathrm{k}_1| - |\mathrm{k}_2| = \frac{2}{3}\frac{\mathrm{C_s}\omega_{\mathrm{pe}}}{\mathrm{v}_{\mathrm{te}}^2} = \frac{2}{3}\left(\frac{\mathrm{m_e}}{\mathrm{m_i}}\right)^{\frac{1}{2}}\left(1 + 3\frac{\mathrm{T_i}}{\mathrm{T_e}}\right)^{1/2}\mathrm{k_{De}}$$

and

$$E_2^-(\mathbf{k}_\mathrm{s},\omega_\mathrm{s}) \cong \frac{e}{m_e}\frac{\omega_s}{2\gamma_s\omega_1^2}k_sE_{11}E_{12}^* \tag{8.7a}$$

where $\mathbf{k}_s \cong 2\mathbf{k}_1$ and

$$\gamma_s = -\left(\frac{\pi}{8}\right)^{\frac{1}{2}} \omega_s \left[\left(\frac{m_e}{m_i}\right)^{\frac{1}{2}} + \left(\frac{T_e}{T_i}\right)^{\frac{3}{2}} \exp\left(-\frac{T_e}{2T_i} - \frac{3}{2}\right)\right]$$

$$\sim -\left(\frac{\pi}{8}\right)^{\frac{1}{2}} \omega_s \left(\frac{m_e}{m_i}\right)^{\frac{1}{2}}$$

for $\frac{T_i}{T_e} \ll 1$.

2. Example 2: Coupling of Langmuir waves and ion acoustic waves, which broadens the Langmuir spectrum.

In (8.6), set $(\mathbf{k}_1, \omega_1)$ and $(\mathbf{k}_2, \omega_2)$ be the Langmuir and ion acoustic mode, respectively. Because $\omega_2 \ll \omega_1$, it requires that $k_\pm \sim -k_1$; thus $|k_2| \sim 2|k_1|$ and $(\mathbf{k}_+, \omega_+)$ is generated by two anti-parallel propagating modes and $(\mathbf{k}_-, \omega_-)$ is generated by two parallel propagating modes. In this situation,

$$\chi_e(k_1, \omega_1) \cong -1 = \chi_e(k_\pm, \omega_\pm) \quad \text{and} \quad \chi_i(k_1, \omega_1) \cong 0 = \chi_i(k_\pm, \omega_\pm);$$

$$\chi_e(k_2, \omega_2) \cong \frac{k_{De}^2}{k_2^2} \cong -\chi_i(k_2, \omega_2);$$

$$\epsilon_L(k_\pm, \omega_\pm) \cong i\left(\frac{\pi}{2}\right)^{\frac{1}{2}} \frac{k_{De}^2}{k_\pm^2} \frac{\omega_\pm}{k_\pm v_{te}} \exp\left(-\frac{\omega_\pm^2}{2k_\pm^2 v_{te}^2}\right);$$

and (8.6) reduces to

$$E_2^\pm(k_\pm, \omega_\pm) \cong \frac{1}{\sqrt{2\pi}} \frac{e}{m_e} \frac{k_1^2 v_{te}}{\omega_1^3} \exp\left(\frac{\omega_1^2}{2k_1^2 v_{te}^2}\right) E_{11} E_{12}^\pm \tag{8.8}$$

So far, we have studied the coupling of two plasma waves, which generates a new plasma mode or drives a fluctuation. On the other hand, the self-coupling of plasma waves, retained on the RHS term of (8.4a), modifies the background distribution.

8.4 Quasi-linear diffusion and equivalent temperature

We now consider a spectral distribution of plasma waves in the background plasma, the total wave field is given by $\boldsymbol{E}_1(\mathbf{r}, t) = \sum_k \boldsymbol{E}_{1k} \exp[i(\mathbf{k} \cdot \mathbf{r} - \omega_k t)] + c.c.$, where $\omega_k = \omega_{kr} + i\gamma_k$, and ω_{kr} and γ_k are the real frequency and damping (or growth) rate of the $(\mathbf{k}, \omega_k)$ mode. Substitute (8.5a) to the

RHS of (8.4a), yields a diffusion equation

$$\frac{\partial f_{a0}}{\partial t} + \mathbf{v}\cdot\nabla f_{a0} = \left(\frac{q_a}{m_a}\right)^2 \nabla_v \cdot \left[\sum_{\mathbf{k}} \frac{2\gamma_k \boldsymbol{E}_{1k}\boldsymbol{E}_{1k}^*}{(\mathbf{k}\cdot\mathbf{v}-\omega_{kr})^2+\gamma_k^2}\right]\cdot\nabla_v f_{a0}$$

$$= \nabla_v \cdot \underline{\underline{\mathbf{D}}}_a \cdot \nabla_v f_{a0} \tag{8.9}$$

where

$$\underline{\underline{\mathbf{D}}}_a(v) = \left(\frac{q_a}{m_a}\right)^2 \sum_k \frac{2\gamma_k \boldsymbol{E}_{1k}\boldsymbol{E}_{1k}^*}{(\mathbf{k}\cdot\mathbf{v}-\omega_{kr})^2+\gamma_k^2}$$

is the diffusion tensor in velocity space.

To simply the illustration, we consider one-dimension case in a uniform plasma, which reduces (8.9) to

$$\frac{\partial}{\partial t} f_{a0} = \frac{\partial}{\partial v} D_a(v) \frac{\partial}{\partial v} f_{a0} \tag{8.10}$$

where

$$D_a(v) = \left(\frac{q_a}{m_a}\right)^2 \sum_k \frac{2\gamma_k |E_{1k}|^2}{(kv-\omega_{kr})^2+\gamma_k^2}$$

is the diffusion coefficient; as shown $|D_i(v)| \ll |D_e(v)|$. In the following, we focus on electron diffusion. $D_e(v)$ is large in the resonance region, around $v = \frac{\omega_{kr}}{k}$, where the diffusion process is called quasi-linear diffusion. It flattens the electron distribution function in the resonance region to form a plateau.

In the bulk region of the distribution, where $v \ll \frac{\omega_{kr}}{k}$, wave and charged particles interact non-resonantly. In the case of Langmuir waves, most of the plasma is in the non-resonance region, thus,

$$D_e(v) \cong \left(\frac{e}{m_e}\right)^2 \sum_k \frac{2\gamma_k |E_{1k}|^2}{\omega_{kr}^2} = \left(\frac{e}{m_e}\right)^2 \frac{d}{dt}\left[\sum_k \frac{|E_{1k}|^2}{\omega_{kr}^2}\right]$$

and (8.10) becomes

$$\frac{\partial}{\partial t} f_{a0} = \frac{1}{2m_e}\left(\frac{d}{dt} T_w\right) \frac{\partial^2}{\partial v^2} f_{a0} \tag{8.11}$$

where

$$T_w = 2\epsilon_0 \sum_k \frac{\omega_{pe}^2 |E_{1k}|^2}{n_0 \omega_{kr}^2}$$

is the time average wave energy per electron.

Multiple $1/2\mathrm{m_e v^2}$ to both sides of (8.11) and integrate both side over the velocity axis, i.e.,

$$\int_{-\infty}^{\infty} dv\, 1/2\mathrm{m_e v^2} \frac{\partial}{\partial \mathrm{t}} \mathrm{f_{a0}} = \int_{-\infty}^{\infty} dv\, 1/2\mathrm{m_e v^2} \frac{1}{2\mathrm{m_e}} \left(\frac{\mathrm{d}}{\mathrm{dt}} \mathrm{T_w} \right) \frac{\partial^2}{\partial \mathrm{v}^2} \mathrm{f_{a0}} \tag{8.11a}$$

With the aids of

$$\int_{-\infty}^{\infty} dv\, 1/2\mathrm{m_e v^2 f_{a0}} = \frac{1}{2} n_0 T_e$$

and

$$\int_{-\infty}^{\infty} dv\, \mathrm{v}^2 \frac{\partial^2}{\partial \mathrm{v}^2} \mathrm{f_{a0}} = 2n_0$$

Eq. (8.11a) becomes

$$\frac{\mathrm{d}}{\mathrm{dt}} \mathrm{T_e} = \frac{\mathrm{d}}{\mathrm{dt}} \mathrm{T_w}$$

which is integrated to be

$$\mathrm{T_{eff}} = \mathrm{T_{e0}} + \mathrm{T}_w$$

where $\mathrm{T_{e0}}$ is the electron temperature of unperturbed background plasma. Let $\mathrm{f_{a0}(v, t)} = u(\mathrm{v}, \mathrm{T_w})$, and with the aid of

$$\frac{\partial}{\partial \mathrm{t}} \mathrm{f_{a0}} = \frac{\partial u}{\partial \mathrm{T_w}} \frac{\mathrm{d}}{\mathrm{dt}} \mathrm{T_w}$$

Eq. (8.11) reduces to a diffusion equation

$$\frac{\partial}{\partial \mathrm{T_w}} u = \frac{1}{2\mathrm{m_e}} \frac{\partial^2}{\partial \mathrm{v}^2} u \tag{8.11b}$$

which is similar to (1.12b) or (7.28) with the notation change: $u \to \phi$, $\mathrm{T_w} \to \mathrm{t}, \mathrm{v} \to \mathrm{x}$, and $\frac{1}{2\mathrm{m_e}} \to \mp\mathrm{D}$ or b. Hence, (8.11b) has the general solution (7.31) or the specific solution (1.12c) for a Gaussian shape initial condition. Since the unperturbed distribution

$$\mathrm{f_{e0}(v, T_w = 0)} = \mathrm{n_0} \sqrt{\frac{\mathrm{m_e}}{2\pi \mathrm{T_{e0}}}} \exp\left(-\frac{\mathrm{m_e v^2}}{2\mathrm{T_{e0}}} \right),$$

is in Gaussian shape, it makes it easy to apply (1.12c) with the substitution of $a = 0$, $\mathrm{b} = (2\mathrm{T_{e0}}/\mathrm{m_e})^{1/2}$, $\mathrm{D} = \mp\frac{1}{2\mathrm{m_e}}$, $\mathrm{E} = 0$, and $\mathrm{t} = \mathrm{T_w}$. The solution of (8.11) is obtained to be

$$\mathrm{f_{e0}(v, T_w) = n_0 \sqrt{\frac{m_e}{2\pi T_{eff}}} \exp\left(-\frac{m_e v^2}{2T_{eff}}\right)} \tag{8.12}$$

As shown, the bulk of the electron plasma has an effective temperature T_{eff}; the temperature increment T_w is proportional to the wave energy density. This is realized that the Langmuir waves introduce quiver motion of non-resonant electrons, the time average kinetic energy of the electron quiver motion in the Langmuir wave fields is

$$\mathrm{KE} = \frac{1}{2} m_e \sum_k \left(\frac{2e|E_{1k}|}{m_e \omega_{kr}}\right)^2 = \mathrm{T_w}.$$

When the spectral energy density of the Langmuir wave is large, it can effectively modify the wave dispersion relations.

8.5 Renormalization of quasilinear diffusion equation–resonance broadening

When plasma (dielectric medium in general) is energized by intense coherence energy sources, instabilities (such as beam-plasma instability excited by electron beam and parametric instabilities excited by electromagnetic wave) are excited to generate broadband plasma waves, leading plasma to a turbulent state. Mode coupling further broadens the wave spectra, which has been discussed in Sec. 8.3. However, in the formulation of mode coupling and quasilinear diffusion, the truncated iterative solution of the Vlasov equation has time secularities in the individual terms of the perturbation expansion. Therefore, a singular perturbation approach or a renormalization procedure is necessary to deal with strong turbulence.

In secular perturbation approach, an unperturbed trajectory:

$$\frac{\mathrm{d}}{\mathrm{dt}}\mathrm{r} = \mathrm{v} \quad \text{and} \quad \frac{\mathrm{d}}{\mathrm{dt}}\mathrm{v} = 0$$

is introduced to convert the Vlasov equation (8.1a) to the form:

$$\frac{d}{dt}\mathrm{f_a} = -\mathbf{E} \cdot \nabla_{\mathrm{v}} \mathrm{f_a},$$

so that it can be integrated along this unperturbed trajectory. However, each expansion term of the response to a time harmonic field contains a singularity (e.g., Eq. (8.5a)), which makes difficulty to justify the truncation of the perturbation expansion. Such a difficulty may be mitigated by introducing a perturbed trajectory, such as

$$\frac{\mathrm{d}}{\mathrm{dt}}\mathrm{r} = \mathrm{v} \quad \text{and} \quad \frac{\mathrm{d}}{\mathrm{dt}}\mathrm{v} = \mathrm{E}, \tag{8.13a}$$

which convert (8.1a) to be

$$\frac{d}{dt}\mathrm{f_a} = 0 \text{ along the trajectory,}$$

so that

$$\mathrm{f_a(v, r, t) = f_a(v_0, r_0, t_0)} \tag{8.13b}$$

where

$$\mathrm{v_0 = v} - \int_{\mathrm{t_0}}^{\mathrm{t}} \mathrm{E}[\mathrm{r}(\tau), \tau]\mathrm{d}\tau \tag{8.13c}$$

and

$$\mathrm{r_0 = r} - \int_{\mathrm{t_0}}^{\mathrm{t}} \mathrm{v}(\tau)\mathrm{d}\tau \tag{8.13d}$$

It seems to resolve the time secularity issue, but it adds difficulty to solve the trajectory equations. The singular perturbation is a good approach when a perturbed trajectory is obtainable. On the other hand, the renormalization approach is to introduce a self-consistent response in the perturbation expansion and identify the iteration patten in the expansion, so that the expansion series converges and can be resumed. In the following, a renormalization procedure to determine the diffusion equation in the turbulent fields is illustrated.

A turbulent field in plasma is modelled as

$$E(\boldsymbol{r}, t) = \sum_k \boldsymbol{E}_{\mathrm{k}} \exp[i(\mathbf{k} \cdot \mathbf{r} - \omega_{\mathrm{k}}\mathrm{t})] + c.c \tag{8.14}$$

It is assumed that each component $\boldsymbol{E}_{\mathrm{k}}$ of the turbulent field carries a random initial phase such that the essemble average of $\boldsymbol{E}_{\mathrm{k}}$ over the initial phase is zero, i.e., $\langle \boldsymbol{E}_{\mathbf{k}} \rangle = 0$, for all k. So far, the Vlasov equation (8.1a) is expanded to include only up to the second order responses, i.e., $\mathrm{f_a = f_{a0} + (f_{a1} + f_{a2} + }c.c.)$, to the first order perturbation field $\mathbf{E}_1 = \boldsymbol{E}_1 + \mathbf{c.c.}$. On the other hand, all of the higher order responses to the turbulent field (8.14) will be included in the following analysis. In other words,

$\mathrm{f_a} = \mathrm{f_{a0}} + (\mathrm{f_{a1}} + \mathrm{f_{a2}} + \mathrm{f_{a3}} \ldots c.c.)$; however, such perturbation expansion will be tedious and a renormalization procedure is applied. The perturbation expansion is regrouped to divide the distribution function into three parts:

$$f = \langle f \rangle + f^{(c)} + \tilde{f},$$

where $\langle f \rangle$, $f^{(c)}$, and $\tilde{f}$ are the average distribution, the phase coherent response to the turbulent field, and the phase incoherent response to the fields induced by the nonlinear wave-wave interaction, respectively. Set $f^{(1)} = f^{(c)} + \tilde{f}$ and expand the spatial dependence of $f^{(1)}$ in a Fourier series

$$f^{(1)}(\mathbf{v}, \mathbf{r}, t) = \sum_k f_k \exp[\mathrm{i}(\mathbf{k} \cdot \mathbf{r} - \omega_\mathrm{r}\mathrm{t})] \tag{8.15}$$

The equation of the average distribution is given to be

$$\left(\frac{\partial}{\partial t} + \mathbf{v} \cdot \nabla\right) \langle f \rangle = -\frac{q_a}{m_a} \nabla_\mathrm{v} \cdot \sum_k \langle E_k^* f_k \rangle \tag{8.16}$$

where $f_k = f_k^{(c)} + \tilde{f}_k$ and

$$\begin{aligned} f_k^{(c)} &= -\frac{q_a}{m_a} \frac{\boldsymbol{E}_\mathrm{k}}{i(\mathbf{k} \cdot \mathbf{v} - \omega_\mathrm{k})} \cdot \nabla_\mathrm{v} \langle f \rangle + \left(\frac{q_a}{m_a}\right)^2 \sum_{k' \neq k} \frac{1}{i(\mathbf{k} \cdot \mathbf{v} - \omega_\mathrm{k})} \\ &\quad \times \nabla_\mathrm{v} \cdot \frac{\boldsymbol{E}_{\mathrm{k}'} \boldsymbol{E}_{\mathrm{k}'}^*}{i[(\mathbf{k} - \mathbf{k}') \cdot \mathbf{v} - (\omega_\mathrm{k} - \omega_{\mathrm{k}'})]} \cdot \nabla_\mathrm{v} f_k^{(c)} \\ &\quad + \left(\frac{q_a}{m_a}\right)^2 \sum_{k' \neq k} \frac{1}{i(\mathbf{k} \cdot \mathbf{v} - \omega_\mathrm{k})} \nabla_\mathrm{v} \cdot \frac{\boldsymbol{E}_{\mathrm{k}'} \boldsymbol{E}_\mathrm{k}}{i[(\mathbf{k} - \mathbf{k}') \cdot \mathbf{v} - (\omega_\mathrm{k} - \omega_{\mathrm{k}'})]} \\ &\quad \cdot \nabla_\mathrm{v} f_{k'}^{(c)*} \end{aligned} \tag{8.17}$$

and

$$\tilde{f}_k = \left(\frac{q_a}{m_a}\right)^2 \sum_{k' \neq k} \frac{1}{i(\mathbf{k} \cdot \mathbf{v} - \omega_\mathrm{k})} \nabla_\mathrm{v} \cdot \frac{\boldsymbol{E}_{\mathrm{k}-\mathrm{k}'} \boldsymbol{E}_{\mathrm{k}'}}{i(\mathbf{k}' \cdot \mathbf{v} - \omega_{\mathrm{k}'})} \cdot \nabla_\mathrm{v} \langle f \rangle + \cdots \tag{8.18}$$

Equation (8.17) can be solved iteratively, and $f_k^{(c)}$ can be expressed in terms of $\langle f \rangle$ including all the higher order terms. The results of the expansion are then resumed, with the following aids:

1. $$\frac{\boldsymbol{E}_{\mathrm{k}}}{i(\mathbf{k}\cdot\mathbf{v}-\omega_{\mathrm{k}})}=\int_0^\infty \boldsymbol{E}_{\mathrm{k}}\exp[-\mathrm{i}(\mathbf{k}\cdot\mathbf{v}-\omega_{\mathrm{k}})\mathrm{t}]\mathrm{dt}$$

2. $$\nabla_{\mathrm{v}}\cdot\frac{\boldsymbol{E}_{\mathrm{k}'}\boldsymbol{E}^*_{\mathrm{k}'}}{i[(\mathbf{k}-\mathbf{k}')\cdot\mathbf{v}-(\omega_{\mathrm{k}}-\omega_{\mathrm{k}'})]}\cdot\nabla_{\mathrm{v}}\int_0^\infty \boldsymbol{E}_{\mathrm{k}}\exp[-\mathrm{i}(\mathbf{k}\cdot\mathbf{v}-\omega_{\mathrm{k}})\mathrm{t}]\mathrm{dt}$$
$$=-\frac{\mathbf{k}\cdot\boldsymbol{E}_{\mathrm{k}'}\boldsymbol{E}^*_{\mathrm{k}'}\cdot\mathbf{k}}{i[(\mathbf{k}-\mathbf{k}')\cdot\mathbf{v}-(\omega_{\mathrm{k}}-\omega_{\mathrm{k}'})]}\int_0^\infty \mathrm{t}^2\boldsymbol{E}_{\mathrm{k}}\exp[-\mathrm{i}(\mathbf{k}\cdot\mathbf{v}-\omega_{\mathrm{k}})\mathrm{t}]\mathrm{dt}$$
$$\sim-\frac{\mathbf{k}\cdot\boldsymbol{E}_{\mathrm{k}'}\boldsymbol{E}^*_{\mathrm{k}'}\cdot\mathbf{k}\tau^2}{i[(\mathbf{k}-\mathbf{k}')\cdot\mathbf{v}-(\omega_{\mathrm{k}}-\omega_{\mathrm{k}'})]}\int_0^\infty \boldsymbol{E}_{\mathrm{k}}\exp[-\mathrm{i}(\mathbf{k}\cdot\mathbf{v}-\omega_{\mathrm{k}})\mathrm{t}]\mathrm{dt}$$

3. $$\sum_{\mathbf{k}'\neq\mathbf{k}}\frac{\mathbf{k}\cdot\boldsymbol{E}_{\mathrm{k}'}\boldsymbol{E}^*_{\mathrm{k}'}\cdot\mathbf{k}\tau^2}{i[(\mathbf{k}-\mathbf{k}')\cdot\mathbf{v}-(\omega_{\mathrm{k}}-\omega_{\mathrm{k}'})]}$$
$$=-\tau^2\int_0^\infty\sum_{\mathbf{k}'\neq\mathbf{k}}\mathbf{k}\cdot\boldsymbol{E}_{\mathrm{k}'}\boldsymbol{E}^*_{\mathrm{k}'}\cdot\mathbf{k}\exp\{-i[(\mathbf{k}'-\mathbf{k})\cdot\mathbf{v}(\omega_{\mathrm{k}'}-\omega_{\mathrm{k}})]\mathrm{t}\}dt$$
$$\sim-\tau^2\mathbf{k}\cdot\underline{\underline{\mathbf{D}}}\cdot\mathbf{k}/\left(\frac{q_a}{m_a}\right)^2$$

4. $$\left\{1-\left(\frac{q_a}{m_a}\right)^2\sum_{k'\neq k}\frac{1}{i(\mathbf{k}\cdot\mathbf{v}-\omega_{\mathrm{k}})}\frac{\mathbf{k}\cdot\boldsymbol{E}_{\mathrm{k}'}\boldsymbol{E}^*_{\mathrm{k}'}\cdot\mathbf{k}\tau^2}{i[(\mathbf{k}-\mathbf{k}')\cdot\mathbf{v}-(\omega_{\mathrm{k}}-\omega_{\mathrm{k}'})]}\right\}^{-1}$$
$$=\frac{i(\mathbf{k}\cdot\mathbf{v}-\omega_{\mathrm{k}})}{i(\mathbf{k}\cdot\mathbf{v}-\omega_{\mathrm{k}})+\tau^2\mathbf{k}\cdot\underline{\underline{\mathbf{D}}}\cdot\mathbf{k}}$$
$$\sim i(\mathbf{k}\cdot\mathbf{v}-\omega_{\mathrm{k}})\int_0^\infty d\tau\exp\left[-i(\mathbf{k}\cdot\mathbf{v}-\omega_{\mathrm{k}})\tau-\frac{1}{3}\mathbf{k}\cdot\underline{\underline{\mathbf{D}}}\cdot\mathbf{k}\tau^3\right]$$

where the renormalized diffusion tensor is given to be

$$\underline{\underline{\mathbf{D}}}=\left(\frac{q_a}{m_a}\right)^2\int_0^\infty d\tau\sum_k\langle\boldsymbol{E}_{\mathbf{k}}\boldsymbol{E}^*_{\mathbf{k}}\rangle\exp\left[-i(\mathbf{k}\cdot\mathbf{v}-\omega_{\mathrm{k}})\tau-\frac{1}{3}\mathbf{k}\cdot\underline{\underline{\mathbf{D}}}\cdot\mathbf{k}\tau^3\right] \tag{8.19}$$

Eq. (8.17) is then resumed to be represented approximately by an integral

$$f_k^{(c)}=-\frac{q_a}{m_a}\int_0^\infty d\tau\exp\left\{-i\left[(\mathbf{k}\cdot\mathbf{v}-\omega_{\mathrm{k}})\tau-\frac{1}{3}\mathbf{k}\cdot\underline{\underline{\mathbf{D}}}\cdot\mathbf{k}\tau^3\right]\right\}\boldsymbol{E}_{\mathrm{k}}\cdot\nabla_{\mathrm{v}}\langle f\rangle \tag{8.20}$$

If only wave-particle resonance interactions are considered, we may substitute (8.20) into (8.16) and neglect (8.18) ascribed to the nonlinear wave-wave interactions. Thus, the renormalized quasi-linear diffusion equation is obtained to be

$$\left(\frac{\partial}{\partial t} + \mathbf{v} \cdot \nabla\right) \langle f \rangle = \nabla_{\mathrm{v}} \cdot \underline{\underline{\mathbf{D}}} \cdot \nabla_{\mathrm{v}} \langle f \rangle \tag{8.21}$$

It shows that the resonance diffusion region broadens as the turbulence level increases. Consequently, the wave-particle interactions become more effective than that based on quasi-linear diffusion.

8.6 Collapse of nonlinear waves

Intense Langmuir waves can be excited in plasma, for example, by the parametric instabilities and by the electron beam-plasma instability. Depend on the operation mode of the driver, the excited Langmuir waves can be periodic waves or localized wave packets. Sections 8.3 to 8.5 show that the nonlinearity of the plasma gives rise to the mode coupling. It broadens the spatial spectrum of the Langmuir waves, which diffuse tail electrons in the velocity distribution via quasi-linear and further resonance broadening mechanisms. On the other hand, Langmuir wave packets may be generated by the localized drivers. In Chap.6, a nonlinear Schrödinger equation (6.8) is derived to describe the propagation of Langmuir wave packets in plasma. This equation is analyzed in Chap.7 in the one-dimensional case for the eigen-state. However, (6.8) can be extended to the multi-dimensional cases. In multi-dimensional systems, the solution of the equation may not be stable; one way to explore the stability conditions is via the "Virial Theorem".

Extend the variance $V(\tau)$ of (7.1f) for 1-dimensional system to n-dimensional system, it becomes

$$V(\tau) = \int |\mathbf{r}|^{\mathbf{2}} |\varphi(\mathbf{r}, \tau)|^2 d\mathbf{r} \tag{8.22}$$

The variance identity becomes

$$\frac{d^2}{d\tau^2} V(\tau) = 4H + (2 - n)\alpha \int |\varphi(\mathbf{r}, \tau)|^4 d\mathbf{r} \tag{8.23}$$

Eq. (8.23) is integrated to be

$$\begin{aligned} V(\tau) &= V(0) + \tau \frac{\mathrm{d}}{\mathrm{d}\tau} V(0) + 2\mathrm{H}\tau^2 \\ &\quad + (2 - n)\alpha \int_0^{\tau} \left[(\tau - s) \int |\varphi(\mathbf{r}, s)|^4 d\mathbf{r} \right] ds \end{aligned} \tag{8.24}$$

where $V(\tau) \geq 0$ for all τ;

$$\frac{\mathrm{d}}{\mathrm{d}\tau}V(0) = i\int |\varphi_0|^2 \mathbf{r}\cdot\nabla\left(ln\frac{\varphi_0^*}{\varphi_0}\right)d\mathbf{r} = 2\int \mathbf{r}\cdot \mathrm{J}_{10}d\mathbf{r};$$

and J_{10} is the initial momentum density defined in Sec. 7.1.1.A. If the RHS can become negative at finite time τ, it suggests that the solution collapses. When $n \geq 2$, the last term on the RHS of (8.24) is zero for $\mathrm{n} = 2$ and negative for $\mathrm{n} = 3$. Therefore, if $\mathrm{H} < 0$, collapse can occur.

In an isotropic plasma, $\nabla^2 = \frac{1}{\mathrm{r}^2}\frac{\mathrm{d}}{\mathrm{dr}}\mathrm{r}^2\frac{\mathrm{d}}{\mathrm{dr}}$, where r is the radial coordinate in the spherical coordinates, (6.8) is re-expressed to be

$$-{}^1\!/_2\frac{1}{\mathrm{r}^2}\frac{\mathrm{d}}{\mathrm{dr}}\mathrm{r}^2\frac{\mathrm{d}}{\mathrm{dr}}\mathrm{E} - \alpha_1|\mathrm{E}|^2\mathrm{E} = \mathrm{i}\frac{\partial}{\partial \mathrm{t}}\mathrm{E} \tag{8.25}$$

Substitute $\mathrm{E} = \mathrm{A(r,t)/r}$ into (8.25), yields

$$-{}^1\!/_2\frac{\mathrm{d}^2}{\mathrm{dr}^2}\mathrm{A} - \frac{\alpha_1|\mathrm{A}|^2\mathrm{A}}{\mathrm{r}^2} = \mathrm{i}\frac{\partial}{\partial \mathrm{t}}\mathrm{A} \tag{8.26}$$

This is an inhomogeneous one-dimensional nonlinear Schrödinger equation, in which the nonlinear potential $\mathrm{V(r,t)} = -\alpha_1|\mathrm{A}|^2/\mathrm{r}^2$ is proportional to the wave intensity ($|\mathrm{A}|^2$) as well as to the location explicitly ($\propto 1/\mathrm{r}^2$).

As the excited waves coalesce into localized wave packets (condensation and nucleation phase), the induced ponderomotive forces push out local plasma to generate density cavities. Because of $1/\mathrm{r}^2$ dependent, the nonlinear wave functions will continue to steepen, and density cavities become deeper and more local (localization phase); the process does not reach a steady state; instead, the steepened nonlinear waves collapse into short wavelength propagating waves (i.e., the steepened nonlinear waves become localized sources to generate new waves). These waves suffer significant Landau damping (dissipation and burnout phase) by the bulk elections in the velocity distribution. On the other hand, the mode coupling process broadens the spatial spectrum of linear Langmuir wave into the long wavelength regime. Those waves, having large phase velocities, facilitate quasilinear and resonance broadening diffusion of the tail electrons in the velocity distribution. After burnout, the remaining waves relax into linear waves, which are then amplified by the source to repeat the cycle. Numerical simulations have illustrated such a nonlinear wave collapse process, the results show that the velocity and density distributions of the plasma are modified significantly by the driven factor of the instability via this turbulent heating process.

Problems

P8.1. In (8.4b) the wave field $E_1 = E_{10}\cos(kx - \omega t)$, which propagates at the phase velocity $\mathrm{v_p} = \frac{\omega}{\mathrm{k}}$.
Find $\mathrm{f_{a1}(x, v, t)}$ around $\mathrm{v_p}$.
[Hint: take Taylor's expansion of $\mathrm{f_{a0}(v)}$ at $\mathrm{v} = \frac{\omega}{\mathrm{k}}$; i.e., $\mathrm{f_{a0}(v)} \cong \mathrm{f_{a0}}\left(\frac{\omega}{\mathrm{k}}\right) + \left(\mathrm{v} - \frac{\omega}{\mathrm{k}}\right)\frac{\partial}{\partial \mathrm{v}}\mathrm{f_{a0}}\left(\frac{\omega}{\mathrm{k}}\right) + \cdots$]

P8.2. The initial condition of the quasi-linear diffusion equation (8.10) is $\mathrm{f_{e0}(v, 0)} = \mathrm{n_0}\sqrt{\frac{\mathrm{m_e}}{2\pi \mathrm{T_e}}}\exp\left(-\frac{\mathrm{m_e v^2}}{2\mathrm{T_e}}\right)$, find $\mathrm{f_{e0}}(\mathrm{v}, t)$ in the two cases:

(1) $\mathrm{D_a(v)} = \mathrm{D_0}$ = constant
(2) $\mathrm{D_a(v)} = \alpha_0/\mathrm{v}$

P8.3. Show that (8.22) indicates that $\int_0^{2\pi} d\varphi \int_0^{\pi} sin\theta d\theta \int_0^{\infty} |E|^2 r^2 dr$ is conserved.

Answers to Problems

Chapter 1

P1.1. Let $\varphi(z,\, t) = \phi(\xi, \eta)$, Eq. (1.1) becomes $-4\mathrm{v}^2\phi_{\xi\eta} = 0$; after integration on η, yields $\phi_\xi = F_\xi(\xi)$. Take another integration on ξ, it leads to $\phi = F(\xi) + G(\eta)$.

P1.2. Take the Laplace transform of Eq. (1.1) and apply integration by part twice on the first term:

$$\int_0^\infty exp(i\omega t)\left(\frac{\partial^2}{\partial t^2} - \mathrm{v}^2\frac{\partial^2}{\partial z^2}\right)U(z,\, t)dt = 0$$

$$= -\frac{\partial}{\partial \mathrm{t}}\mathrm{U}(\mathrm{z},\, 0) + \mathrm{i}\omega\mathrm{U}(\mathrm{z},\, 0) - \omega^2 L(z,\omega) - \mathrm{v}^2\frac{\partial^2}{\partial \mathrm{z}^2}L(z,\omega)$$

Hence,

$$\frac{\partial^2}{\partial \mathrm{z}^2}\mathrm{L}(\mathrm{z},\, \omega) + \frac{\omega^2}{\mathrm{v}^2}\mathrm{L}(\mathrm{z},\omega) = \frac{\mathrm{i}\,\omega}{\mathrm{v}^2}\left[\mathrm{U}(\mathrm{z},\, 0) + \frac{\mathrm{i}}{\omega}\frac{\partial}{\partial \mathrm{t}}\mathrm{U}(\mathrm{z},\, 0)\right]$$

P.1.3.

$$\int_{-\infty}^{\infty}\mathrm{U}(0,\, \mathrm{t})\mathrm{e}^{\mathrm{i}\omega \mathrm{t}}\mathrm{dt} = \frac{1}{4\pi}\left[\iint_{-\infty}^{\infty}\mathrm{F}(\omega')\exp[\mathrm{i}(\omega - \omega')\mathrm{t}]\mathrm{dt}\mathrm{d}\omega' + \iint_{-\infty}^{\infty}\mathrm{F}^*(\omega')\exp[\mathrm{i}(\omega + \omega')\mathrm{t}]\mathrm{dt}\,\mathrm{d}\omega'\right]$$

$$= \frac{1}{2}[\mathrm{F}(\omega) + \mathrm{F}^*(-\omega)]$$

$$\int_{-\infty}^{\infty}\frac{\partial \mathrm{U}(0,\, \mathrm{t})}{\partial \mathrm{z}}\mathrm{e}^{\mathrm{i}\omega \mathrm{t}}\mathrm{dt}$$

$$= \frac{i}{4\pi}\left[\iint_{-\infty}^{\infty} k(\omega)F(\omega')\exp[i(\omega-\omega')t]dt\,d\omega'\right.$$

$$\left.-\iint_{-\infty}^{\infty} k(\omega)F^*(\omega')\exp[i(\omega+\omega')t]dt\,d\omega'\right]$$

$$= \frac{i}{2}k(\omega)[F(\omega)-F^*(-\omega)]$$

where $k(-\omega) = k(\omega)$ is assumed. Hence,

$$F(\omega) = \int_{-\infty}^{\infty}\left[U(0,\,t) - \frac{i}{k(\omega)}\frac{\partial U(0,\,t)}{\partial z}\right]e^{i\omega t}\,dt$$

P1.4.

$$u(x,\,t) = \mathrm{Re}\left\{\frac{1}{\sqrt{1+i\beta t}}\left[\exp\left\{-\frac{(x-v_g t)^2}{2L^2(1+i\beta t)}\right.\right.\right.$$

$$\left.+i\left[k_0x-\omega(k_0)t-\frac{\pi}{2}\right]\right\}+\exp\left\{-\frac{(x+v_g t)^2}{2L^2(1+i\beta t)}\right.$$

$$\left.\left.\left.-i\left[k_0x+\omega(k_0)t-\frac{\pi}{2}\right]\right\}\right]\right\}$$

where "Re" stands for the "real part of"; $\beta = \frac{2\alpha\omega_0}{L^2}$; $v_g = 2\omega_0\alpha k_0$; and $\omega(k_0) = \omega_0(1+\alpha\,k_0^2)$.

P1.5. In case (b), $\phi(x,\,t) = \frac{1}{\sqrt{1+2t}}\exp\left(-\frac{x^2}{1+2t}\right)$; it leads to

$$\frac{\partial\phi}{\partial t} = \left[-(1+2t)^{-\frac{3}{2}} + 2x^2(1+2t)^{-\frac{5}{2}}\right]\exp\left(-\frac{x^2}{1+2t}\right)$$

$$\frac{\partial\phi}{\partial x} = \left[-2x(1+2t)^{-\frac{3}{2}}\right]\exp\left(-\frac{x^2}{1+2t}\right)$$

$$\frac{\partial^2\phi}{\partial x^2} = 2\left[-(1+2t)^{-\frac{3}{2}} + 2x^2(1+2t)^{-\frac{5}{2}}\right]\exp\left(-\frac{x^2}{1+2t}\right)$$

Hence,

$$\frac{\partial\phi}{\partial t} - 0.5\frac{\partial^2\phi}{\partial x^2} = 0$$

In case (c), $\phi(x,\,t) = \frac{1}{\sqrt{1+i2t}}\exp\left(-\frac{x^2}{1+i2t}\right)$

$$\frac{\partial\phi}{\partial t} = i\left[-(1+i2t)^{-\frac{3}{2}} + 2x^2(1+i2t)^{-\frac{5}{2}}\right]\exp\left(-\frac{x^2}{1+i2t}\right)$$

$$\frac{\partial \phi}{\partial \mathrm{x}} = \left[-2x(1+i2t)^{-\frac{3}{2}}\right] \exp\left(-\frac{\mathrm{x}^2}{1+\mathrm{i2t}}\right)$$

$$\frac{\partial^2 \phi}{\partial \mathrm{x}^2} = 2\left[-(1+i2t)^{-\frac{3}{2}} + 2x^2(1+i2t)^{-\frac{5}{2}}\right] \exp\left(-\frac{\mathrm{x}^2}{1+\mathrm{i2t}}\right)$$

Hence,

$$\frac{\partial \phi}{\partial \mathrm{t}} - \mathrm{i}0.5\,\frac{\partial^2 \phi}{\partial \mathrm{x}^2} = 0$$

P1.6. (1) $\mathrm{A} = 0 = \mathrm{B}$:

$$\mathrm{H}_1(\mathrm{z},\, \omega) = \mathrm{e}^{\mathrm{ik}(\omega)\mathrm{z}} = \mathrm{e}^{\mathrm{i}\omega\frac{\mathrm{z}}{\mathrm{c}}} \quad \text{and} \quad h_1(z,\, t) = \delta\left(\mathrm{t} - \frac{\mathrm{z}}{\mathrm{c}}\right)$$

$$\hat{E}(z,\, t) = \int_{-\infty}^{\infty} h_1(z,\, t')\hat{E}(0,\, t-t')\, dt'$$

$$= \int_{-\infty}^{\infty} \delta\left(\mathrm{t}' - \frac{\mathrm{z}}{\mathrm{c}}\right) \tilde{A}(0,\, t-t')e^{-i\omega(t-t')}u(t-t')dt'$$

$$= \tilde{A}\left(0,\, t - \frac{\mathrm{z}}{\mathrm{c}}\right) e^{-i\omega\left(t-\frac{\mathrm{z}}{\mathrm{c}}\right)} u\left(t - \frac{\mathrm{z}}{\mathrm{c}}\right)$$

(2) $\mathrm{A} > 0$ and $\mathrm{B} = 0$:

$$\mathrm{H}_2(\mathrm{z},\, \omega) = \mathrm{e}^{\mathrm{ik}(\omega)\mathrm{z}} = \mathrm{e}^{\mathrm{i}\omega\frac{\mathrm{z}}{\mathrm{c}}}\, \mathrm{e}^{-\mathrm{A}\frac{\mathrm{z}}{\mathrm{c}}}$$

$$h_2(z,\, t) = \delta\left(\mathrm{t} - \frac{\mathrm{z}}{\mathrm{c}}\right) \mathrm{e}^{-\mathrm{A}\frac{\mathrm{z}}{\mathrm{c}}}$$

$$\hat{E}(z,\, t) = \tilde{A}\left(0,\, t - \frac{\mathrm{z}}{\mathrm{c}}\right) \mathrm{e}^{-\mathrm{A}\frac{\mathrm{z}}{\mathrm{c}}} e^{-i\omega\left(t-\frac{\mathrm{z}}{\mathrm{c}}\right)} u\left(t - \frac{\mathrm{z}}{\mathrm{c}}\right)$$

(3) $\mathrm{A} = \mathrm{B} = \alpha > 0$:

$$h_3(z,\, t) = \delta\left(\mathrm{t} - \frac{\mathrm{z}}{\mathrm{c}}\right) \mathrm{e}^{-\alpha\frac{\mathrm{z}}{\mathrm{c}}} + \alpha\frac{\mathrm{z}}{\mathrm{c}} \frac{\mathrm{I}_1\left(\alpha\sqrt{\mathrm{t}^2 - \left(\frac{\mathrm{z}}{\mathrm{c}}\right)^2}\right)}{\sqrt{\mathrm{t}^2 - \left(\frac{\mathrm{z}}{\mathrm{c}}\right)^2}}$$

$$\times\, \mathrm{e}^{-\alpha \mathrm{t}} \mathrm{u}\left(\mathrm{t} - \frac{\mathrm{z}}{\mathrm{c}}\right)$$

$$\hat{E}(z,\, t) = \tilde{A}\left(0,\, t - \frac{\mathrm{z}}{\mathrm{c}}\right) \mathrm{e}^{-\alpha\frac{\mathrm{z}}{\mathrm{c}}} e^{-i\omega\left(t-\frac{\mathrm{z}}{\mathrm{c}}\right)}\, u\left(t - \frac{\mathrm{z}}{\mathrm{c}}\right)$$

$$+\, \alpha\frac{\mathrm{z}}{\mathrm{c}} \int_{-\infty}^{\infty} \frac{\mathrm{I}_1\left(\alpha\sqrt{t'^2 - \left(\frac{\mathrm{z}}{\mathrm{c}}\right)^2}\right)}{\sqrt{t'^2 - \left(\frac{\mathrm{z}}{\mathrm{c}}\right)^2}} \mathrm{e}^{-\alpha \mathrm{t}'}\, \mathrm{u}\left(t' - \frac{\mathrm{z}}{\mathrm{c}}\right)$$

$$\times \tilde{A}(0, t-t')e^{-i\omega(t-t')}\, u(t-t')dt' = [\tilde{A}\left(0, t-\frac{z}{c}\right)$$

$$\times e^{-\alpha\frac{z}{c}} e^{-i\omega\left(t-\frac{z}{c}\right)} + \alpha\frac{z}{c} e^{-i\omega t} \int_{-\infty}^{t} \frac{I_1\left(\alpha\sqrt{t'^2-\left(\frac{z}{c}\right)^2}\right)}{\sqrt{t'^2-\left(\frac{z}{c}\right)^2}}$$

$$\times e^{i(\omega+i\alpha)t'} \tilde{A}(0, t-t')dt']u\left(t-\frac{z}{c}\right) \sim \tilde{A}\left(0, t-\frac{z}{c}\right)$$

$$\times e^{-\alpha\frac{z}{c}} e^{-i\omega\left(t-\frac{z}{c}\right)}$$

$$\times \left[1 + \alpha\frac{z}{c}\int_{-\infty}^{t} \frac{I_1\left(\alpha\sqrt{t'^2-\left(\frac{z}{c}\right)^2}\right)}{\sqrt{t'^2-\left(\frac{z}{c}\right)^2}} dt'\right] u\left(t-\frac{z}{c}\right)$$

P1.7. $A(z, t) = \frac{1}{2}A_0\left[erfc\left(\frac{t}{\sqrt{-i2\beta_2 z}}\right) - erfc\left(\frac{t+t_0}{\sqrt{-i2\beta_2 z}}\right)\right]$

Chapter 2

P2.1. $\int k(z)dz = -k_0\int tanh(z/L)dz = -k_0 L\,\ell n\left[2cosh\left(\frac{z}{L}\right)\right]$

$$E(z) = -\frac{E_0}{\tanh\left(\frac{z}{L}\right)}\exp\left\{-ik_0 L\,\ell n\left[2cosh\left(\frac{z}{L}\right)\right]\right\}$$

P2.2. $\frac{d\mathbf{r}}{dt} = \nabla_{\mathbf{k}}\omega = \mathbf{v}_g = \mathbf{k}c^2/\omega;\ \frac{d\mathbf{k}}{dt} = -\nabla\omega = -\hat{\mathbf{z}}\frac{1}{2\omega}\frac{\partial}{\partial z}\omega_p^2(z)$

P2.3. Ray trajectory equations:

$$\frac{dx}{dt} = k_x c^2/\omega;\ \frac{dz}{dt} = \frac{k_z c^2}{\omega};\ \frac{dk_x}{dt} = 0;\ \frac{dk_z}{dt} = -\frac{\omega_{p0}^2}{2\omega z_0}$$

Hence, $k_x = k_{x0}$, $x = \frac{k_{x0}c^2}{\omega}t$, $k_z = k_{z0} - \frac{\omega_{p0}^2}{2\omega z_0}t = k_{z0} - \frac{1}{2}k_{x0}\alpha^2\frac{x}{z_0}$, and $\frac{dz}{dx} = \frac{k_z}{k_{x0}} = \alpha - \frac{1}{2}\alpha^2\frac{x}{z_0}$

It leads to $\frac{z}{z_0} = 1 - \left(1 - \frac{1}{2}\alpha\frac{x}{z_0}\right)^2$.

P2.4. The ray equations are

$$\frac{dx}{dt} = \frac{k_x c^2}{\omega};\ \frac{dy}{dt} = \frac{k_y c^2}{\omega};\ \frac{dk_x}{dt} = 0;\ \frac{dk_y}{dt} = -\frac{\omega_{p0}^2 y}{y_0^2\omega};$$

which subject to the initial conditions $x(0) = x_0$, $y(0) = 0$, and $\mathbf{k}(0) = \hat{x}k_{x0} + \hat{y}k_{y0}$, where $k_0 = (k_{x0}^2 + k_{y0}^2)^{1/2} = \omega/c$. First,

$k_x = const. = k_{x0}$; thus we have $k_y = k_{y0}[1 - (\omega_{p0}/k_{y0}c)^2(y/y_0)^2]^{1/2} = k_{y0}[1 - (y/y_0)^2]^{1/2}$, where $(\omega_{p0}/k_{yo}c) = 1$, and $x = x_0 + k_{x0}c^2t/\omega$. The trajectory equation on the x-y plane is then derived to be

$$\frac{dy}{dx} = \frac{k_y}{k_x} = \left(\frac{k_{y0}}{k_{x0}}\right)\left[1 - \left(\frac{\omega_{p0}}{k_{y0}c}\right)^2\left(\frac{y}{y_0}\right)^2\right]^{\frac{1}{2}}.$$

It is integrated to obtain ray trajectory y(x) on the x-y plane to be

$$y(x) = \left(\frac{k_{y0}}{k_{x0}}\right)\kappa_x^{-1}\, sin[\kappa_x(x - x_0)],$$

where $\kappa_x = \omega_{p0}/k_{x0}cy_0$. The ray trajectory (4.23) is plotted in Fig. P2.1. As shown, this ray is guided by the lens to propagate along its central axis.

P2.5. $d^2y/dt^2 = (c^2/\omega)dk_y/dt = -\left(\frac{\omega_{p0}^2c^2}{y_0^2\omega^2}\right)y = -\omega_B^2 y$

$$\Rightarrow \omega_B = (\omega_{p0}c/\omega y_0) \quad \text{and} \quad y(t) = y_0 \sin\omega_B t$$

P2.6. $\frac{dz}{dt} = \frac{k_zc^2}{\omega} = \frac{k_0c^2}{\omega}e^{-\frac{z}{2L}}; \ \frac{dk_z}{dt} = -\left(\frac{\omega_0^2}{2\omega L}\right)e^{\frac{z}{L}};$

$$z = 2L\ell n\left(1 + \frac{k_0c^2}{2\omega L}t\right) \quad \text{and} \quad k = k_0/\left(\frac{1 + k_0c^2}{2\omega L}t\right)$$

P2.7. $\frac{dt(s)}{ds} = t\, and\, \frac{dx(s)}{ds} = -2x$; it leads to $t = e^s\, \text{and}\, x_0 = xt^2$; and $\frac{d}{ds}u = t\frac{d}{dt}u$. Thus (2.30) becomes

$$\frac{1}{u^2}\frac{du}{dt} = \frac{1}{t}$$

It is integrated to be

$$u(x, t) = \frac{u(x_0, t_0)}{1 - u(x_0, t_0)ln(\frac{t}{t_0})} = \frac{xt^2}{1 - xt^2ln(t)}$$

P2.8.

$$\frac{\partial u}{\partial t} + \alpha x^2\frac{\partial u}{\partial x} = \beta u^2$$

$$u(x, t) = \frac{\varphi[x(1 + \alpha xt)]}{1 - \beta t\varphi[x(1 + \alpha xt)]}$$

$$= \frac{cos[kx(1 + \alpha xt)]}{1 - \beta t cos[kx(1 + \alpha xt)]}$$

P2.9. u(x, t)

$$= \frac{(1 + 2a\alpha xt + b\alpha t) \pm \sqrt{(1 + 2a\alpha xt + b\alpha t)^2 - 4a\alpha^2t^2(ax^2 + bx + c)}}{2a\alpha^2t^2}$$

Chapter 3

P3.1. One can apply superposition principle in linear system. Hence, (3.4) and (3.6) for linear polarization case are extended to

$$\begin{aligned}\boldsymbol{E}(\mathrm{z},\,\mathrm{t}>0) &= A_+\left[\hat{\mathbf{x}}\cos(\mathrm{k_0 z}-\omega t)-\hat{\mathbf{y}}\sin(\mathrm{k_0 z}-\omega t)\right]\\ &\quad + \mathrm{A}_-\hat{\mathbf{x}}\cos(\mathrm{k_0 z}+\omega\mathrm{t})-\hat{\mathbf{y}}\sin(\mathrm{k_0 z}+\omega\mathrm{t})\\ \boldsymbol{H}(\mathrm{z},\,\mathrm{t}>0) &= \left(\frac{\mathrm{k_0}}{\mu_0\omega}\right)A_+\left[\hat{\mathbf{y}}\cos(\mathrm{k_0 z}-\omega\mathrm{t})\right.\\ &\quad \left.+\hat{\mathbf{x}}\sin(\mathrm{k_0 z}-\omega t)\right]-A_-\left[\hat{\mathbf{y}}\cos(\mathrm{k_0 z}+\omega\mathrm{t})\right.\\ &\quad \left.+\hat{\mathbf{x}}\sin(\mathrm{k_0 z}+\omega\mathrm{t})\right]+\boldsymbol{H}_{\mathrm{w}}(\mathrm{z})\end{aligned}$$

where $A_+ = (1+\omega_0/\omega)E_0/2$ and $A_- = (1-\omega_0/\omega)\,E_0/2$;

$$\omega = \sqrt{\omega_{\mathrm{p}}^2+\omega_0^2}; \quad \text{and} \quad \boldsymbol{H}_{\mathrm{w}}(\mathrm{z}) = \left(\frac{\epsilon_0 c\omega_p^2}{\omega^2}\right)E_0(\hat{\mathbf{y}}\cos\mathrm{k_0 z}+\hat{\mathbf{x}}\sin\mathrm{k_0 z})$$

P3.2. (1) For $-\ell < \mathrm{z} < \ell$, we have $(\partial_{\mathrm{z}}^2+\mathrm{k}^2)E(z)=0$, where $\mathrm{k}=\frac{\omega}{\mathrm{c}}$; thus,

$$E(z) = \mathrm{A}\,\exp(\mathrm{ikz})+\mathrm{B}\exp(-\mathrm{ikz}).$$

For $\ell < \mathrm{z} < \ell+\mathrm{d}$, we have $(\partial_{\mathrm{z}}^2+\mathrm{k}_1^2)E(z)=0$, where $k_1=\xi k$; thus,

$$E(z) = \mathrm{C}\exp[\mathrm{ik_1(z}-\ell)]+\mathrm{D}\exp[\mathrm{ik_1(z}-\ell)]$$

(2) At $z = -\ell$;

$$\begin{aligned}E(-\ell^-) &= \exp(-\mathrm{i}\beta\mathrm{L})[\mathrm{C}\exp(\mathrm{ik_1}d)+\mathrm{D}\exp(-\mathrm{ik_1}d)] = E(-\ell^+)\\ &= \mathrm{A}\,\exp(-\mathrm{ik}\ell)+\mathrm{B}\,\exp(\mathrm{ik}\ell)\\ \partial_{\mathrm{z}}E(-\ell^-) &= \mathrm{ik_1}[\mathrm{C}\,\exp(-2\mathrm{ik_1}\ell)-\mathrm{D}\,\exp(2i\mathrm{k_1}\ell)]\\ &= \partial_{\mathrm{z}}E(-\ell^+) = \mathrm{ik}[\mathrm{A}\,\exp(-\mathrm{ik}\ell)-\mathrm{B}\exp(\mathrm{ik}\ell)]\end{aligned}$$

At $z = \ell$;

$$\begin{aligned}E(\ell^-) &= \mathrm{A}\,\exp(\mathrm{ik}\ell)+\mathrm{B}\,\exp(-\mathrm{ik}\ell) = E(\ell^+) = \mathrm{C}+\mathrm{D}\\ \partial_{\mathrm{z}}E(\ell^-) &= \mathrm{i}k[\mathrm{A}\exp(\mathrm{ik}\ell)-\mathrm{B}\,\exp(-\mathrm{ik}\ell)]\\ &= \partial_{\mathrm{z}}E(\ell^+) = \mathrm{ik_1(C-D)}\end{aligned}$$

It leads to the equations:

$$\mathrm{A}\exp(-\mathrm{i}k\ell) + \mathrm{B}\exp(\mathrm{i}k\ell)$$
$$= \exp(-\mathrm{i}\beta\mathrm{L})[\mathrm{C}\exp(\mathrm{i}k_1 d) + \mathrm{D}\exp(-\mathrm{i}k_1 d)]$$

$$\mathrm{A}\exp(-\mathrm{i}k\ell) - \mathrm{B}\exp(-\mathrm{i}k\ell)$$
$$= \left(\frac{k_1}{k}\right)\exp(-\mathrm{i}\beta\mathrm{L})[\mathrm{C}\exp(\mathrm{i}k_1 d) - \mathrm{D}\exp(-\mathrm{i}k_1 d)]$$

$$\mathrm{A}\exp(\mathrm{i}k\ell) + \mathrm{B}\exp(-\mathrm{i}k\ell) = \mathrm{C} + \mathrm{D}$$

$$\mathrm{A}\exp(\mathrm{i}k\ell) - \mathrm{B}\exp(-\mathrm{i}kl) = \left(\frac{k_1}{k}\right)(\mathrm{C} - \mathrm{D})$$

The ratio of the first two equation (to remove $\exp(-\mathrm{i}\beta\mathrm{L})$ factor) together with the last two equations are solved to obtain (P3.5). Substitute (P3.5) into the first one, (P3.6) is derived.

(3) $E_0\exp(-\mathrm{i}k_0 z) = \exp(-\mathrm{i}\beta\mathrm{L})E_0\exp[-\mathrm{i}k_0(z+\mathrm{L})]$ It leads to $\exp(-\mathrm{i}(\beta+k_0)\mathrm{L}) = 1$

$$\Rightarrow (\beta + k_0)L = 2p\pi \Rightarrow \beta + k_0 = \frac{2\pi p}{L}.$$

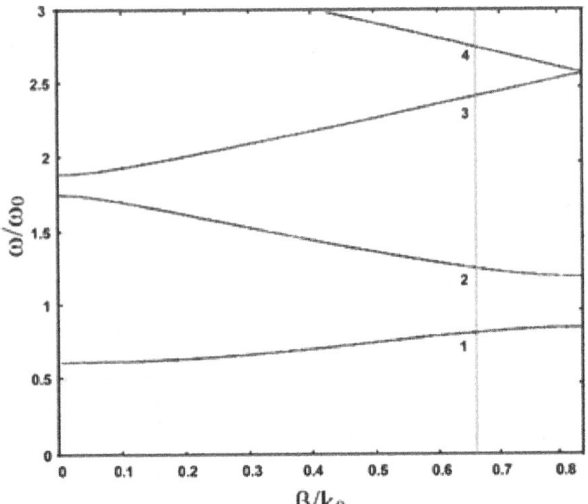

(4) Dispersion relation $\omega(\beta)$ for the case $\mathrm{L} = 0.6\lambda_0$, and $\omega_{p0} = 1.2\omega_0$, where λ_0 and ω_0 are the wavelength and angular frequency of an

initial reference wave in free space. The vertical line is at

$$\beta/k_0 = 2\pi/k_0L - 1 = 0.667$$

and its intersecting points with the dispersion curves determine the frequencies of the Floquet modes converted from the initial reference wave after interacting with the suddenly created periodic plasma.

Chapter 4

P4.1. $\dfrac{\mathrm{d}^2q}{\mathrm{d}t^2} + \Omega^2 q = -h\cos\Omega t,$

The ansatz $q(\mathrm{t}) = a(t)\sin\theta + b(t)\cos\theta$ is assumed with $\theta = \theta_0 + \Omega t$; thus,

$$\frac{\mathrm{d}^2a}{\mathrm{d}t^2}\sin\theta + \frac{\mathrm{d}^2b}{\mathrm{d}t^2}\cos\theta + 2\frac{\mathrm{d}a}{\mathrm{d}t}\Omega\cos\theta - 2\frac{\mathrm{d}b}{\mathrm{d}t}\Omega\sin\theta$$
$$= -h\,(\cos\theta_0\cos\theta + \sin\theta_0\sin\theta)$$

It leads to $\frac{\mathrm{d^2a}}{\mathrm{dt^2}} - 2\frac{\mathrm{db}}{\mathrm{dt}}\Omega = -\mathrm{h}\sin\theta_0$ and $\frac{\mathrm{d^2b}}{\mathrm{dt^2}} + 2\frac{\mathrm{da}}{\mathrm{dt}}\Omega = -\mathrm{h}\cos\theta_0$, which become

$$\frac{\mathrm{d}a}{\mathrm{d}t} = -\frac{h}{2\Omega}\cos\theta_0 \quad \text{and} \quad \frac{\mathrm{d}b}{\mathrm{d}t} = \frac{h}{2\Omega}\sin\theta_0$$

Hence, $a = a_0 - \left(\frac{h}{2\Omega}\cos\theta_0\right)t$ and $b = b_0 + \left(\frac{h}{2\Omega\sin\theta_0}\right)t$.
The initial conditions: $q(0) = q_0$ and $\dot{q}(0) = 0$ lead to $a_0 = q_0\sin\theta_0$ and $b_0 = q_0\cos\theta_0$; then

$$q(\mathrm{t}) = q_0 cos\Omega t - \left(\frac{h}{2\Omega}\right) t\, sin\Omega t.$$

P4.2. $\dfrac{dq}{\mathrm{d}t} = -(\omega\sin\theta + \alpha\cos\theta)ae^{-\alpha t} - \nu b\sin(\nu t + \phi)$

$$\frac{\mathrm{d}^2q}{\mathrm{d}t^2} = \left[-(\omega^2 - \alpha^2)\cos\theta + 2\alpha\omega\sin\theta\right]ae^{-\alpha t} - \nu^2 b\cos(\nu t + \phi)$$

Hence, (P4.1) leads to

$$\Omega^2 - \omega^2 + \alpha^2 - \gamma\alpha = 0;\ 2\alpha\omega - \gamma\omega = 0;\ (\Omega^2 - \nu^2)sin\,\phi$$
$$= -\gamma\nu\cos\phi;\ \text{and}$$
$$b = -h/[(\Omega^2 - \nu^2)cos\phi - \gamma\nu sin\phi]$$
$$\alpha = \frac{\gamma}{2};\ \omega = \sqrt{\Omega^2 - \frac{\gamma^2}{4}};$$

$$\phi = -tan^{-1}\left(\frac{\gamma\nu}{\Omega^2 - \nu^2}\right),$$

$$b = -\frac{\mathrm{h}\left(\Omega^2 - \nu^2\right) sec\phi}{(\Omega^2 - \nu^2)^2 + (\gamma\nu)^2}, \theta = \theta_0 + \omega t, \text{ and}$$

$$acos\theta_0 + bcos\phi = q_0\ and\ a(\alpha cos\theta_0 + \omega sin\theta_0) + b\nu sin\phi$$
$$= 0 \text{ for } a\, and\, \theta_0$$

P4.3. $\frac{\mathrm{d}^2\mathrm{q}}{\mathrm{dt}^2} + \Omega^2\mathrm{q} - 2\alpha^2\mathrm{q}^3 = 0$

Let $\mathrm{t} = \mathrm{a}\tau$; and $\mathrm{a}^2\Omega^2 = 1 + \mathrm{k}^2$ and $\mathrm{a}^2\alpha^2 = \mathrm{k}^2$, where $\mathrm{a}^2 = \frac{1}{\Omega^2 - \alpha^2}$ and $\mathrm{k}^2 = \frac{\alpha^2}{\Omega^2 - \alpha^2}$.

P4.4. $\frac{d^2q}{dt^2} + D\frac{dq}{dt} + Eq + Gq^3 = 0$; $U(q) = \frac{E}{2}q^2 + \frac{G}{4}q^4$

$$\frac{dU(q)}{dq} = Eq + Gq^3 = 0 = Eq\left(1 - \frac{G}{|E|}q^2\right)$$

The three equilibria are: $\mathrm{q}_1 = 0, \mathrm{q}_2 = -\sqrt{\frac{|\mathrm{E}|}{\mathrm{G}}}$, and $\mathrm{q}_3 = \sqrt{\frac{|\mathrm{E}|}{\mathrm{G}}}$.

P4.5. The Ansatz $q(t) = a(t)\sin\theta$ is chosen, where $\mathrm{d}\theta/\mathrm{dt} = \Omega(\mathrm{t})$, defining equations for a and Ω follow from

$$\Omega^2 = \omega_0^2\left[1 - \frac{\epsilon^2}{2}\left(1 - \frac{1}{4}\beta a^2\right)\left(1 - \frac{3}{4}\beta a^2\right)\right]$$

$$a(t) = \frac{2}{\sqrt{\beta}}\left[1 + \left(\frac{4}{\beta a_0^2} - 1\right)e^{-\epsilon\omega_0 t}\right]^{-\frac{1}{2}}$$

where $a_0 = a(0) = \frac{\mathrm{v}_0}{\Omega_0}$;

$$\Omega_0 = \Omega(0) = \omega_0\left[1 - \frac{\epsilon^2}{2}\left(1 - \frac{1}{4}\beta\left(\frac{\mathrm{v}_0}{\Omega_0}\right)^2\right)\left(1 - \frac{3}{4}\beta\left(\frac{\mathrm{v}_0}{\Omega_0}\right)^2\right)\right]^{1/2}.$$

P4.6. Poincaré section plots on the phase plane in the three cases: (a) $\mathrm{h} = 0$; (b) $\mathrm{h} = 0.05$; and (c) $\mathrm{h} = 0.2$.

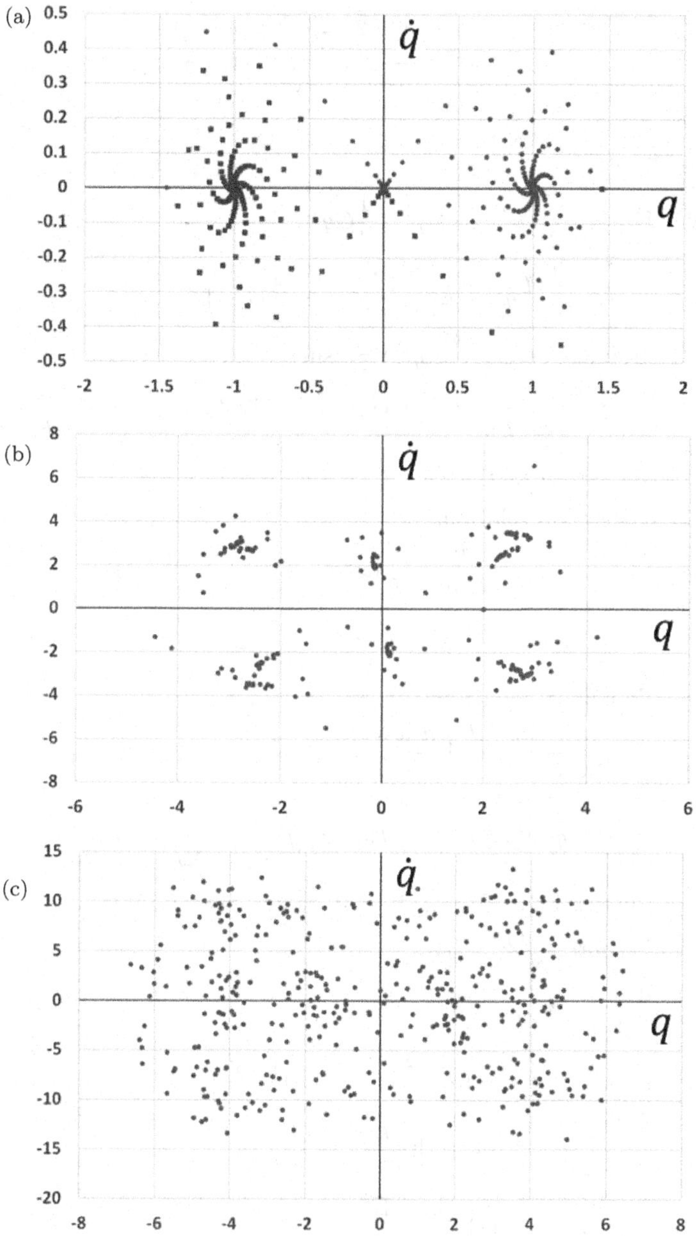
(a)
$\dot{q}$
q
(b)
$\dot{q}$
q
(c)
$\dot{q}$
q

P4.7.

$$x(t) = \frac{2\sqrt{\alpha}}{\sqrt{\gamma + 3\beta\omega^2}} cos(\omega t + \phi)$$

Chapter 5

P5.1. (1) $V(q) = \frac{1}{2}Eq^2 + \frac{1}{3}Fq^3 + \frac{1}{4}Gq^4$

$$L = \frac{1}{2}\left(\dot{q}^2 - Eq^2 - \frac{2}{3}Fq^3 - \frac{1}{2}Gq^4\right)$$

(2) Assume the ansatz: $q(\mathrm{t}) = a\sin\theta$

$$\mathcal{L} = \frac{1}{2\pi}\int_0^{2\pi} L(q, \dot{q}, \mathrm{t})d\theta = \frac{1}{4}\left[a^2\omega^2 - Ea^2\left(1 + \frac{3G}{8E}a^2\right)\right]$$

$$= \mathcal{L}(a, \omega)$$

(3) Dispersion relation:

$$\frac{\partial \mathcal{L}}{\partial a} = \frac{1}{2}\left(\omega^2 a - Ea - \frac{3}{4}\mathrm{G}a^3\right) = 0$$

$$\omega = \pm\sqrt{E + \frac{3}{4}\mathrm{G}a^2}$$

Transport equation:

$$\frac{d}{dt}\frac{\partial \mathcal{L}}{\partial \omega} = \frac{d}{dt}\left(\frac{a^2\omega}{2}\right) = 0$$

$$q(t) = a(t)sin\theta(t) = a_0\sqrt{\frac{\omega_0}{\omega}}sin\left(\int_0^t \omega(t')dt' + \theta_0\right)$$

where $\omega_0 = \omega(0) = \sqrt{E + \frac{3}{4}\mathrm{G}a_0^2}$ and $a_0 sin\theta_0 = q_0$ and $a_0 cos\theta_0 = \frac{v_0}{\omega_0}(note{:}\frac{da(0)}{dt} = 0)$.

P5.2. (1) $\frac{\mathrm{d}^2\mathrm{E}}{\mathrm{dz}^2} + \mathrm{k}^2(\mathrm{z})\mathrm{E} = 0$

where $\mathrm{k}^2(\mathrm{z}) = \left[1 - \frac{\omega_0^2}{\omega^2}\left(1 - \mathrm{e}^{-\frac{\mathrm{z}}{L}}\right)\right] k_0^2$.

(2)

$$\frac{1}{2}\frac{d}{\mathrm{dz}}\dot{E}^2 + \mathrm{k}^2(\mathrm{z})\mathrm{E}\frac{dE}{dz} = 0$$

$$\frac{1}{2}\frac{d}{\mathrm{dz}}(\dot{E}^2 + \mathrm{k}^2(\mathrm{z})\mathrm{E}^2) = \frac{\mathrm{E}^2}{2}\frac{d}{dz}\mathrm{k}^2(\mathrm{z})$$

$$L = \frac{1}{2}(\dot{E}^2 - \mathrm{k}^2(\mathrm{z})\mathrm{E}^2)$$

(3) Assume the ansatz: $q(\mathrm{z}) = a \sin\theta$, where $\frac{d\theta}{dz} = \kappa$.

$$\mathcal{L}(a,\,\kappa) = \frac{1}{2\pi}\int_0^{2\pi} L(q, \dot{q}, \mathrm{t})d\theta = \frac{1}{4}[a^2\kappa^2 - \mathrm{k}^2(\mathrm{z})a^2]$$

Dispersion relation: $\frac{\partial\mathcal{L}}{\partial a} = \frac{1}{2}\left[\kappa^2 a - \mathrm{k}^2\,(\mathrm{z})\,a\right] = 0$

$$\Rightarrow \kappa = \pm\, k(z).$$

Transport equation: $\frac{d}{dt}\frac{\partial\mathcal{L}}{\partial\kappa} = \frac{d}{dt}\left(\frac{a^2\kappa}{2}\right) = 0$

$$\Rightarrow \quad a\,(z) = a\,(0)\sqrt{\frac{k_0}{k}} = a_0\sqrt{\frac{k_0}{k}}$$

$$E(z) = a(z) sin\theta(z) = a_0\sqrt{\frac{\mathrm{k}_0}{\mathrm{k}}} sin\left(\int_0^z \mathrm{k}(z')dz' + \theta_0\right),$$

where $a_0 = \frac{E_0}{sin\theta_0}$.

(4) It is the same.

P5.3. (1) $H(E, \dot{E}, \mathrm{z}) = \frac{1}{2}(\dot{E}^2 + \mathrm{k}^2(\mathrm{z})\mathrm{E}^2)$

(2) $\frac{\partial H}{\partial \dot{E}} = \dot{E}$ and $\frac{\partial H}{\partial E} = \mathrm{k}^2(\mathrm{z})E = -\frac{d}{dz}\dot{E} \Rightarrow \frac{\mathrm{d}^2\mathrm{E}}{\mathrm{dz}^2} + \mathrm{k}^2(\mathrm{z})\mathrm{E} = 0$

P5.4. (1) $\ddot{x} - \frac{\lambda x\dot{x}^2}{1+\lambda x^2} + \frac{\alpha}{1+\lambda x^2}x = 0$

(2) $x(t) = A\,cos(\omega t + \varphi)$, *with* $\omega = \frac{\alpha}{\sqrt{1+\lambda A^2}}$.

P5.5. (1) $L = \frac{1}{\dot{x} + kx^2}$

(2) $x(t) = \frac{2t}{kt^2 - E}$, E is a constant.

P5.6. (1) $\ddot{\mathrm{x}} + 3\mathrm{k}\mathrm{x}\dot{\mathrm{x}} + \mathrm{k}^2\mathrm{x}^3 + \omega^2\mathrm{x} = 0$

(2) $\ddot{y} + \omega^2 y =$ constant

(3) $x(t) = \frac{\omega A sin(\omega t + \phi)}{1 - kA cos(\omega t + \phi)}$

P5.7. $\omega \dfrac{\partial S}{\partial \mathcal{H}_s} = 1$ and $\frac{dS}{dt} = 0$

$$\frac{\partial S}{\partial \mathcal{H}_s} = \frac{2}{\pi\alpha}\int_0^{q_{2s}} \frac{dq}{\sqrt{(q^2 - q_{1s}^2)(q^2 - q_{2s}^2)}} = \frac{2}{\pi\alpha}\frac{K(\beta_s)}{q_{1s}} = \frac{1}{\omega}$$

$$\Rightarrow \omega = \frac{\pi\alpha q_{1s}}{2K(\beta_s)}$$

$$\frac{dS}{dt} = 0 \Rightarrow \frac{d\mathcal{H}_s}{dt} = 0 \Rightarrow H \cong \mathcal{H}_s = \mathcal{H}_s(0) = H(0) = \frac{\mathrm{v}_0^2}{2}$$

Hence,

$$q(t) = q_2 \,\mathrm{sn}(\mathcal{R}\int_0^{\mathrm{t}} \omega(\mathrm{t}')\mathrm{dt}')$$

$$\dot{q}(0) = q_2\mathcal{R}\omega = \mathrm{v}_0 \Rightarrow \mathcal{R} = \frac{2K(\beta_s)}{\pi}$$

Chapter 6

P6.1. For bounded states, $\varphi_n(\xi,\tau) \sim \varphi_n e^{-\sqrt{2}\alpha_n|\xi|} e^{i\alpha_n^2\tau}$

For unbounded states, $\varphi(\xi,\, \tau) \sim \varphi_0 e^{i\sqrt{2}k|\xi|} e^{-ik^2\tau}$

P6.2. $$-{}^1\!/\!_2\varphi_{\ell^*}\frac{\partial^2}{\partial\xi_1^2}\varphi_\ell - \alpha_1\left|\varphi_\ell\right|^4 = \mathrm{i}\varphi_{\ell^*}\frac{\partial}{\partial\tau_1}\varphi_\ell$$

$$-{}^1\!/\!_2\varphi_\ell\frac{\partial^2}{\partial\xi_1^2}\,\varphi_\ell^* - \alpha_1\left|\varphi_\ell\right|^4 = -\mathrm{i}\varphi_\ell\frac{\partial}{\partial\tau_1}\varphi_\ell^*$$

$$i\frac{\partial}{\partial\tau_1}\left|\varphi_\ell\right|^2 = -\frac{1}{2}\left(\varphi_\ell^*\frac{\partial^2}{\partial\xi_1^2}\varphi_\ell - \varphi_\ell\frac{\partial^2}{\partial\xi_1^2}\varphi_\ell^*\right)$$

$$= -\frac{1}{2}\frac{\partial^2}{\partial\xi_1}\left[\varphi_\ell^{*2}\frac{\partial}{\partial\xi_1}\left(\frac{\varphi_\ell}{\varphi_\ell^*}\right)\right]$$

Hence, $$\frac{\mathrm{d}}{\mathrm{d}\tau_1}\int_{-\infty}^{\infty}\left|\varphi_\ell\right|^2 d\xi_1 = 0.$$

P6.3. Let $x = \eta - 6v\tau$

$$\frac{\partial}{\partial\tau}\tilde{\phi}(\eta,\,\tau) = \frac{\partial}{\partial\tau}\phi(x,\,\tau) - 6v\frac{\partial}{\partial \mathrm{x}}\phi(x,\,\tau)$$

$$\frac{\partial}{\partial\eta}\tilde{\phi}(\eta,\,\tau) = \frac{\partial}{\partial \mathrm{x}}\phi(x,\,\tau)\ and\ \frac{\partial^3}{\partial\eta^3}\tilde{\phi}(\eta,\,\tau) = \frac{\partial^3}{\partial \mathrm{x}^3}\phi(x,\,\tau)$$

$$\frac{\partial}{\partial\tau}\tilde{\phi} + \frac{\partial^3}{\partial\eta^3}\tilde{\phi} + 6\tilde{\phi}\frac{\partial}{\partial\eta}\tilde{\phi} = \frac{\partial}{\partial\tau}\phi - 6v\frac{\partial}{\partial \mathrm{x}}\phi + \frac{\partial^3}{\partial \mathrm{x}^3}\phi + 6\phi\frac{\partial}{\partial \mathrm{x}}\phi$$

$$+ 6v\frac{\partial}{\partial \mathrm{x}}\phi = \frac{\partial}{\partial\tau}\phi(x,\tau) + \frac{\partial^3}{\partial \mathrm{x}^3}\phi(x,\tau)$$

$$+ 6\phi(x,\tau)\frac{\partial}{\partial \mathrm{x}}\phi(x,\tau) = 0$$

P6.4. $\phi_1(\eta,\tau) = \dfrac{\mathrm{a}\eta + \mathrm{b}}{\mathrm{a}\tau + 1}$

Chapter 7

P7.1. Set $\phi(\xi) = \mathrm{Asech(kx)}; it\, is\, found\, that\, E = -\frac{1}{2}k^2.$

P7.2. $\mathrm{z} = \sqrt{2}\mathrm{e}^{-|\xi|}$

$$\frac{\partial}{\partial\xi} \to -\sqrt{2}\mathrm{e}^{-|\xi|}\frac{|\xi|}{\xi}\frac{\partial}{\partial z} and \frac{\partial^2}{\partial\xi^2} \to z[1 - 2\delta(\xi)]\frac{\partial}{\partial z} + z^2\frac{\partial^2}{\partial z^2}$$

Hence, $\frac{\partial^2}{\partial \mathrm{z}^2}\mathrm{y} + \frac{1}{\mathrm{z}}\frac{\partial}{\partial \mathrm{z}}\mathrm{y} + \left(1 - \frac{\nu^2}{\mathrm{z}^2}\right)\mathrm{y} = 0$ for $\xi \neq 0$.

P7.3.

$$i\frac{\partial}{\partial\tau}\left(\frac{\partial\varphi^*}{\partial\xi}\frac{\partial\varphi}{\partial\xi}\right) = \frac{\partial\varphi^*}{\partial\xi}\frac{\partial}{\partial\xi}\left(-{}^1\!/\!{}_2\frac{\partial^2}{\partial\xi^2}\varphi - \alpha|\varphi|^2\varphi\right)$$

$$-\frac{\partial\varphi}{\partial\xi}\frac{\partial}{\partial\xi}\left(-{}^1\!/\!{}_2\frac{\partial^2}{\partial\xi^2}\varphi^* - \alpha|\varphi|^2\varphi^*\right)$$

$$= -{}^1\!/\!{}_2\frac{\partial}{\partial\xi}\left(\frac{\partial\varphi^*}{\partial\xi}\frac{\partial^2}{\partial\xi^2}\varphi - \frac{\partial\varphi}{\partial\xi}\frac{\partial^2}{\partial\xi^2}\varphi^*\right)$$

$$-\alpha\frac{\partial}{\partial\xi}|\varphi|^2\left(\varphi\frac{\partial\varphi^*}{\partial\xi} - \varphi^*\frac{\partial\varphi}{\partial\xi}\right)$$

$$\mathrm{i}\frac{\partial}{\partial\tau}|\varphi|^4 = 2|\varphi|^2\left[\varphi^*\left(-{}^1\!/\!{}_2\frac{\partial^2}{\partial\xi^2}\varphi - \alpha|\varphi|^2\varphi\right)\right.$$

$$\left. -\varphi\left(-{}^1\!/\!{}_2\frac{\partial^2}{\partial\xi^2}\varphi^* - \alpha|\varphi|^2\varphi^*\right)\right]$$

$$= -|\varphi|^2\left[\varphi^*\frac{\partial^2}{\partial\xi^2}\varphi - \varphi\frac{\partial^2}{\partial\xi^2}\varphi^*\right] = |\varphi|^2\frac{\partial}{\partial\xi}$$
$$\times\left(\varphi\frac{\partial\varphi^*}{\partial\xi} - \varphi^*\frac{\partial\varphi}{\partial\xi}\right)$$

$$\frac{\mathrm{d}}{\mathrm{d}\tau}H = \frac{i}{4}\int_{-\infty}^{\infty}\frac{\partial}{\partial\xi}\left[\left(\frac{\partial\varphi^*}{\partial\xi}\frac{\partial^2}{\partial\xi^2}\varphi - \frac{\partial\varphi}{\partial\xi}\frac{\partial^2}{\partial\xi^2}\varphi^*\right)\right.$$
$$\left. + 2\alpha|\varphi|^2\left(\varphi\frac{\partial\varphi^*}{\partial\xi} - \varphi^*\frac{\partial\varphi}{\partial\xi}\right)\right]d\xi = 0$$

P7.4. $u(x\,t) = u_0\,cos(kx - \omega t)$, where $\omega = 1 + k^2$.

P7.5. Set $u(x,\,t) = u(x - \mathrm{v}\,t) = u(\xi)$, $(P7.1)$ becomes

$$(1 - \mathrm{v}^2)u_{\xi\xi} = \sin u$$

Multiply u_ξ to both sides and integrate both sides over ξ, lead to

$$\frac{du}{sin\frac{u}{2}} = \pm\frac{2}{\sqrt{1-\mathrm{v}^2}}d\xi$$

It is then integrated to obtain

$$u(\xi) = 4tan^{-1}\left[tan\left(\frac{u_0}{4}\right)\exp\left(\pm\frac{\xi - \xi_0}{\sqrt{1-\mathrm{v}^2}}\right)\right],$$

where $u_0 = u(\xi_0) = u(x_0,\,0)$ and $\xi_0 = x_0$.

$$u(x,\,t) = 4tan^{-1}\left[tan\left(\frac{u_0}{4}\right)\exp\left(\pm\frac{\mathrm{x} - \mathrm{x}_0 - \mathrm{v}t}{\sqrt{1-\mathrm{v}^2}}\right)\right].$$

P7.6. Introduce the transform $\mathrm{s} = \tanh(\mathrm{x})$, (7.22a) becomes

$$\frac{d}{ds}(1 - s^2)\frac{d}{ds}U + \left[2 + \frac{E}{(1 - s^2)}\right]U = 0$$

Hence, $\ell = 1$ and $m^2 = -E = 1$; it leads to $\kappa_1 = 1$. The eigenfunction, which is proportional to the associated Legendre polynomial P_1^1, is normalized to be

$$\varphi_1(x,0) = -\frac{1}{\sqrt{2}}\,sech(x)$$

Then,

$$c_1(0) = \lim_{x\to\infty} \varphi_1(x,0)e^{\kappa_1 x} = \lim_{x\to\infty}\left[-\frac{1}{\sqrt{2}}\,sech(x)e^{x}\right] = -\sqrt{2}$$

and

$$F(x;t) = c_1^2(0)e^{8\kappa_1^3 t}e^{-\kappa_1 x} = 2e^{8t-x}$$

The GLM linear integral equation becomes

$$K(x,\,y;t) + 2\,e^{8t-(x+y)} + 2\int_x^\infty K(x,\,z;t)e^{8t-(y+z)}dz = 0.$$

It is solved to obtain

$$K(x,\,x;t) = -2\,\frac{e^{8t-2x}}{(1+e^{8t-2x})}$$

Substitute $K(x,\,x;t)$ into (7.19a), a one-soliton solution is obtained to be:

$$\phi(x,t) = 2sech^2(x-4t)$$

It satisfies the initial condition: $\phi(x,\,0) = 2sech^2(x)$.

P7.7. $\phi(\eta,\tau) = 48\dfrac{3+4\cosh(4\eta-64\tau)+\cosh(8\eta-512\tau)}{[3\cosh(2\eta-224\tau)+\cosh(6\eta-288\tau)]^2}$

Chapter 8

P8.1.
$$\frac{\partial \mathrm{f_{a1}}}{\partial \mathrm{t}} + \mathbf{v}\cdot\nabla \mathrm{f_{a1}} = -\left(\frac{\mathrm{q_a}}{\mathrm{m_a}}\right)\boldsymbol{E}_1\cdot\nabla_{\mathbf{v}}\mathrm{f_{a0}}$$
$$\sim -\left(\frac{\mathrm{q_a}}{\mathrm{m_a}}\right)\frac{\partial}{\partial \mathrm{v}}\mathrm{f_{a0}}\Big(\frac{\omega}{\mathrm{k}}\Big)\,\mathbf{E}_1$$

Substitute $\mathrm{f_{a1}(x,\,v,\,t)} = \mathrm{f}_{a10}(\mathrm{v})sin(kx-\omega t)$ into above equation, yields

$$\mathrm{f_{a10}(v)} \sim -\frac{\left(\frac{\mathrm{q_a}}{\mathrm{m_a}}\right)\frac{\partial}{\partial \mathrm{v}}\mathrm{f_{a0}}(\frac{\omega}{\mathrm{k}})\,\mathrm{E_{10}}}{\mathrm{kv}-\omega}$$
$$= \frac{\frac{\mathrm{n_0}}{\sqrt{2\pi}}\left(\frac{\mathrm{q_a}}{\mathrm{m_a}}\right)\left(\frac{\mathrm{m_a}}{\mathrm{T_a}}\right)^{\frac{3}{2}}\frac{\omega}{\mathrm{k}}\exp(-\frac{\mathrm{m_a}\omega^2}{2\mathrm{T_a k^2}})}{\mathrm{kv}-\omega}$$

P8.2. (1) $\frac{\partial}{\partial \mathrm{t}}\mathrm{f_{a0}} = \mathrm{D_0}\frac{\partial^2}{\partial \mathrm{v}^2}\mathrm{f_{a0}}$; this is similar as (1.12b) with

$$D = \mp D_0,\ E = 0,\ b^2 = \frac{2T_e}{m_e},\ and\, x \to \mathrm{v}.$$

Thus, $f_{a0} = \mathrm{n_0}\sqrt{\frac{\mathrm{m_e}}{2\pi \mathrm{T_e}}}\frac{1}{\sqrt{1+2\frac{\mathrm{m_e D_0}}{\mathrm{T}_e}t}}\exp\left[-\frac{\mathrm{m_e v^2}}{2\mathrm{T_e}\left(1+2\frac{\mathrm{m_e D_0}}{\mathrm{T}_e}t\right)}\right]$

(2) $\frac{\partial}{\partial t}\mathrm{f_{a0}} = 4\alpha_0\sqrt{\mathrm{x}}\frac{\partial^2}{\partial \mathrm{x}^2}\mathrm{f_{a0}}$; where $\mathrm{x} = \mathrm{v}^2$.

This is solved numerically.

P8.3. $$-{}^1\!/_2\mathrm{E}^*\frac{1}{\mathrm{r}^2}\frac{\mathrm{d}}{\mathrm{dr}}\mathrm{r}^2\frac{\mathrm{d}}{\mathrm{dr}}\mathrm{E} - \alpha_1|E|^4 = \mathrm{i}\mathrm{E}^*\frac{\partial}{\partial \mathrm{t}}\mathrm{E}$$

$$-{}^1\!/_2\mathrm{E}\frac{1}{\mathrm{r}^2}\frac{\mathrm{d}}{\mathrm{dr}}\mathrm{r}^2\frac{\mathrm{d}}{\mathrm{dr}}\mathrm{E}^* - \alpha_1|E|^4 = -\mathrm{i}\mathrm{E}\frac{\partial}{\partial \mathrm{t}}\mathrm{E}^*$$

$$i\frac{\partial}{\partial \mathrm{t}}|E|^2 = -{}^1\!/_2\left(\frac{E^*}{r^2}\frac{d}{dr}r^2\frac{d}{dr}E - \frac{E}{r^2}\frac{d}{dr}r^2\frac{d}{dr}E^*\right)$$

Let $\mathrm{x} = 1/r \Rightarrow \frac{d}{dr} \to -\frac{1}{r^2}\frac{d}{dx}$; thus,

$$i\frac{\partial}{\partial \mathrm{t}}|E|^2 = -{}^1\!/_2\frac{1}{r^4}\left(\mathrm{E}^*\frac{d^2}{dx^2}E - \mathrm{E}\frac{d^2}{dx^2}\mathrm{E}^*\right)$$

$$= -{}^1\!/_2\frac{1}{r^4}\frac{\partial}{\partial x}\left[\mathrm{E}^{*2}\frac{\partial}{\partial x}\left(\frac{\mathrm{E}}{\mathrm{E}^*}\right)\right]$$

$$\int_0^\infty \frac{1}{r^4}\frac{\partial}{\partial x}\left[\mathrm{E}^{*2}\frac{\partial}{\partial x}\left(\frac{\mathrm{E}}{\mathrm{E}^*}\right)\right]r^2dr$$

$$= \int_0^\infty \frac{\partial}{\partial x}\left[\mathrm{E}^{*2}\frac{\partial}{\partial x}\left(\frac{\mathrm{E}}{\mathrm{E}^*}\right)\right]dx = 0$$

$$\Rightarrow \frac{\mathrm{d}}{\mathrm{dt}}\int_0^{2\pi}d\varphi\int_0^\pi sin\theta d\theta\int_0^\infty |E|^2\, r^2dr = 0$$

Bibliography

Ablowitz, M. J. and P. A. Clarkson, *Solitons, Nonlinear Evolution Equations and Inverse Scattering*, London Mathematical Society lecture Notes 149, Cambridge University press, Cambridge, 1991.

Ablowitz, M., D. Kaup, A. Newell, and H. Segur. "The inverse scattering transform — fourier analysis for nonlinear problems", *Stud. Appl. Math.*, **53**, 249–315, 1974.

Brauer, K., *The Korteweg-de Vries Equation: History, Exact Solutions, and Graphical Representation*, University of Osnabrück, 2000.

Budden, K. G., *Radio Waves in the Ionosphere*, Cambridge University Press, Cambridge, 1961.

Campbell, G. A. and R. M. Foster, *Fourier Integrals for Practical Applications*, Van Nostrand, New York, 1948.

Colliander, J., M. Keel, G. Staffilani, H. Takaoka, T. Tao, "Almost conservation laws and global rough solutions to a nonlinear Schrödinger equation", *Math. Res. Lett.*, **9**, 659–682, 2002.

Davidson, R. C., *Methods in Nonlinear Plasma Theory*, Academic Press, New York, 1972.

de Bouard, A. and A. Debussche, "On the effect of a noise on the solutions of supercritical Schrödinger equation", *Prob. Theory and Rel. Fields*, **123**, 76–96, 2002.

Dodd, R. K., J. C. Eilbeck, J. D. Gibbon, and H. C. Morris, *Solitons and nonlinear wave equations,* Academic Press, London, 1982.

Drazin, P. G. and R. S. Johnson, *Solitons: an Introduction,* Cambridge University Press, Cambridge, 1988.

Dupree, T. H., "A perturbation theory for strong plasma turbulence", *Phys. Fluids*, **9**, 1773–1782, 1966, doi: 10.1063/1.1761932.

Faith, J., S. P. Kuo, J. Huang, and G. Schmidt, "Precipitation of magnetospheric electrons caused by relativistic effect enhanced chaotic motion in the whistler wave fields", *J. Geophy. Res.*, **102**(A5), 9631–9638, 1997.

Gardner, S., J. M. Greene, M. D. Kruskal, and R. M. Miura, "Method for solving the Korteweg–deVries equation", *Phys. Rev. Lett.*, **19**(19), 1095–1097, doi: 10.1103/PhysRevLett.19.1095, 1967.

Ginzburg, V. L., *The Propagation of Electromagnetic Waves in Plasmas*, 2nd ed., Pergamon Press, New York, 1970.

Griffiths, Graham. W. and W. E. Schiesser, "Linear and nonlinear waves", *Scholarpedia*, **4**(7), 4308, 2009.

Griffiths, G. W. and W. E. Schiesser, *Traveling Wave Solutions of Partial Differential Equations: Numerical and Analytical Methods with Matlab and Maple*, Academic Press, 2011.

Kalluri, D. K., *Eletromagnetics of Time Varying Complex Media, Frequency and Polarization Transformer*, 2nd ed. Boca Raton, FL: CRC Press, 2010.

Kaufman, A. N. and L. Stenflo, "Upper-hybrid solitons", *Phys. Scr.*, **11**, 269, doi: 10.1088/0031-8949/11/5/005, 1975.

Koelink, E., *Scattering Theory*, Radboud Universiteit, 2008.

Korteweg, D. J. and G. de Vries, "On the change of form of long waves advancing in a rectangular channel and on a new type of long stationary waves", *Phil. Mag.*, **39**, 422–443, 1895.

Kuo, S., "Frequency up-conversion of microwave pulse in a rapidly growing plasma", *Phys. Rev. Lett.*, **65**(8), 1000–1003, 1990.

Kuo, S., "On the nonlinear plasma waves in the high-frequency (HF) wave heating of the ionosphere", *IEEE Trans. Plasma Sci.*, **42**(4), 1000–1005, 2014; doi: 10.1109/TPS.2014.2306834.

Kuo, S., *Plasma Physics in Active Wave Ionosphere Interaction*, World Scientific, Singapore, 2018. ISBN: 978-981-3232-12-9 (hardcover); https://doi.org/10.1142/10767; ISBN: 978-981-3232-14-3 (ebook).

Kuo, S., "On the propagation of a circularly polarized electromagnetic wave through a temporal interface between air and a magneto plasma", *IEEE Trans. Plasma Sci.*, **47**(10), 4599–4605, 2019; doi: 10.1109/TPS.2019.2939489.

Kuo, S. and B. Watkins, "Nonlinear upper hybrid waves generated in ionospheric HF heating experiments at HAARP", *IEEE Trans. Plasma Sci.*, **47**(12), 5334–5338, 2019; doi:10.1109/TPS.2019.2950752.

Lax, P., "Integrals of nonlinear equations of evolution and solitary waves", *Comm. Pure Applied Math.*, **21**(5), 467–490, 1968; doi:10.1002/cpa.3160210503.

Leibovich, S. and A. Richard Seebass, eds., *Nonlinear Waves*, Cornell University Press, Ithaca and London, 1974.

Marchant, T. R. and N. F. Smyth, "Soliton interaction for the extended Korteweg-de Vries equation", *IMA J. of Applied Mathematics*, **56**, 157–176, 1996.

Marinca, Vasile and Nicolae Heri anu, "Explicit and exact solutions to cubic Duffing and double-well Duffing equations", *Mathematical and Computer Modelling*, **53**, 604–609, 2011.

Miura, R. M., "Korteweg-deVries equation and generalizations. I. A remarkable explicit nonlinear transformation", *J. Math. Phys.*, **9**, 1202–1204, 1968.

Miura, R. M., C. S. Gardener, and M. D. Kruskal, "Korteweg-deVries equation and generalizations. II. Existence of conservation laws and constants of motion", *J. Math. Phys.*, **9**, 1204–1209, 1968.

Novikov, S., S. V. Manakov, L. P. Pitaevskii, and V. E. Zakharov, *Theory of Solitons,* Consultants Bureau, New York, 1984.

Orfanidis, S. J., *Electromagnetic Waves and Antennas*, Rutgers ECE, 1999.

Robinson, P. A., "Nonlinear wave collapse and strong turbulence", *Rev. Modern Phys.*, **69**(2), 507–574, 1997.

Sagdeev, R. Z. and A. A. Galeev, *Nonlinear Plasma Theory,* T. M. O'Neil and D. L. Book, eds., Benjamin Reading, Massachusetts, 1969.

Schmidt, G., *Physics of High Temperature Plasmas*, 2nd Edit., Academic Press, New York, 1979.

Seadawy, A. R., "Exact solutions of a two-dimensional nonlinear Schrödinger equation", *Appl. Math. Lett.*, **25**, 687–691, 2012.

Stenflo, L., "Upper-hybrid wave collapse", *Phys. Rev. Lett.*, **48**(20), 1441, 1982.

Taha, T. R. and M. J. Ablowitz, "Analytical and numerical aspects of certain nonlinear evolution equations. III. Numerical, Korteweg-de Vries equation", *J. Comp. Phys.*, **55**, 231–253, 1984.

Vainshtein, L. A., "Propagation of pulses", *Usp. Fiz. Nauk*, **118**, 339–367 (*Sov. Phys.-Usp.,* **19**, 189–205), 1976.

Vlasov, A. A., "On the kinetic theory of an assembly of particles with collective interaction", *J. Phys.* (U.S.S.R.), **9**, 25, 1945.

Whitham, G. B., *Linear and Nonlinear Waves*, John Wiley Sons, 1999. Print/ Online ISBN: 9780471359425/9781118032954; DOI:10.1002/9781118032954.

Zabusky, N. J. and M. D. Kruskal, "Interaction of solitons in a collisionless plasma and the recurrence of initial states", *Phys. Rev. Lett.*, **15**, 240–243, 1965.

Zakharov, V. E., "Collapse of Langmuir waves", *Zh. Eksp. Teor. Fiz.*, **62**, 1745–1759, 1972; *Soviet Phys. JETP*, **35**(5), 908–914, 1972.

Index

Printed in the United States
by Baker & Taylor Publisher Services